SCIENCE AT THE FRONTIER

Volume I

by
ADDISON GREENWOOD
with Marcia F. Bartusiak, Barbara A. Burke, and Edward Edelson
for the
NATIONAL ACADEMY OF SCIENCES

NATIONAL ACADEMY PRESS
Washington, D.C.   1992

NATIONAL ACADEMY PRESS • 2101 Constitution Ave., N.W. • Washington, D.C. 20418

The National Academy of Sciences is a private, nonprofit, self-perpetuating society of distinguished scholars engaged in scientific and engineering research, dedicated to the furtherance of science and technology and to their use for the general welfare. Upon the authority of the charter granted to it by the Congress in 1863, the Academy has a mandate that requires it to advise the federal government on scientific and technical matters. Dr. Frank Press is president of the National Academy of Sciences.

The National Academy of Engineering was established in 1964, under the charter of the National Academy of Sciences, as a parallel organization of outstanding engineers. It is autonomous in its administration and in the selection of its members, sharing with the National Academy of Sciences the responsibility for advising the federal government. The National Academy of Engineering also sponsors engineering programs aimed at meeting national needs, encourages education and research, and recognizes the superior achievements of engineers. Dr. Robert M. White is president of the National Academy of Engineering.

The Institute of Medicine was established in 1970 by the National Academy of Sciences to secure the services of eminent members of appropriate professions in the examination of policy matters pertaining to the health of the public. The Institute acts under the responsibility given to the National Academy of Sciences by its congressional charter to be an adviser to the federal government and, upon its own initiative, to identify issues of medical care, research, and education. Dr. Kenneth I. Shine is president of the Institute of Medicine.

The National Research Council was organized by the National Academy of Sciences in 1916 to associate the broad community of science and technology with the Academy's purposes of furthering knowledge and advising the federal government. Functioning in accordance with general policies determined by the Academy, the Council has become the principal operating agency of both the National Academy of Sciences and the National Academy of Engineering in providing services to the government, the public, and the scientific and engineering communities. The Council is administered jointly by both Academies and the Institute of Medicine. Dr. Frank Press and Dr. Robert M. White are chairman and vice chairman, respectively, of the National Research Council.

Support for this project was provided by the National Research Council's Basic Science Fund and the National Science Foundation.

International Standard Book Number 0-309-04592-4
International Standard Serial Number 1065-3082

Cover: View of Earth as photographed from the Apollo 17 spacecraft (courtesy of NASA). Magnetic resonance image of a patient's head indicating the shape and location of a tumor (reprinted, by permission, from H.E. Cline et al., 1990. *Journal of Computer Assisted Tomography* 14(6):1037-1045). A young star in the Orion Nebula as shown in a photograph made with the Wide-Field/Planetary camera on the Hubble Space Telescope (courtesy of NASA). Magnified fragment of the Mandelbrot set (courtesy of J.H. Hubbard, Cornell University).

Printed in the United States of America

Foreword

Stephen Jay Gould once remarked that "great scientists have an instinct for the fruitful and the doable." In science, as in most human activities, taste matters—selecting topics that truly advance a field by virtue of new theories, new techniques, or the ineluctable power of new ideas, that break new ground rather than walking over old trails.

This volume testifies by example to the power of "the fruitful and the doable." Whether suggesting the deep insights offered by the mathematics of dynamical systems, discussing the distribution of galaxies, analyzing the intricate and fleeting events that enable photosynthesis and hence life, or outlining computational approaches to sorting out the elaborate complexities of air pollution, each of the chapters in this volume affirms the richness, opportunities, and frontiers of contemporary science.

And each affirms the talent and breadth of the generation now on the verge of assuming the leadership of American science. The work presented in this volume is in almost every instance that of young scientists, 45 years old and under, each already highly distinguished in her or his field. These scientists presented their work at an annual symposium, the Frontiers of Science, organized by the National Academy of Sciences. This remarkable—and to those able to attend it, exhilarating—event brings together the very best young scientists. It is not only the speakers who are carefully selected by their peers, but also the audience. The scientists invited to be in the audience already have accomplished much and have been recognized for their achievements, by the conferring of such highly distinguished awards as the Waterman, and Packard or Sloan fellowships. Further, several of those who have participated in the Frontiers symposia have gone on to win Nobel prizes, MacArthur "genius" awards, or the Fields Medal, perhaps the premier award in core mathematics, or have been elected to membership in the National Academy of Sciences.

The Frontiers of Science symposia were first suggested by Roy Schwitters, now director of the Superconducting Super Collider Laboratory. Their purpose is to enable leading young scientists to present and discuss their work with their peers in other fields, at a level intelligible across fields. The rule is that, for example, a molecular biologist should understand a talk given by a geologist, and vice versa. The first Frontiers symposium, held in 1989, was made possible through the generous support of the National Science Foundation and the Sloan Foundation, as well as by provision of National Academy of Sciences' institutional funds. Since then, the National Science Foundation and the Academy have continued to provide support for the Frontiers symposia. About 100 scientists have attended each of the three symposia held so far—to listen, and to share ideas, techniques, and their excitement over several days and nights.

It is these people who are the immediate audience for this volume, *Science at the Frontier*. At the same time, many of us involved in the Frontiers symposia have come to believe that the work presented in these discussions—indeed representing the frontiers of contemporary science—should be made available to a wider, not necessarily scientific audience that nevertheless may share with the Frontiers participants a passion for currency in the exploration of nature in all its forms. Translating that goal into a book accessible to a wide audience was not easy and required not only the work of talented science writers but also the help, and patience, of the participants whose work is reported in this volume, to assure that their work was fairly and accurately presented.

We are now at work on similar volumes for the 1991 and 1992 Frontiers symposia. Individually and as a set, these volumes will provide to the larger public a view of the dynamics and achievements of the American science that they so generously support and that has so richly benefited and will continue to benefit us all.

Frank Press, President
National Academy of Sciences

Contents

SCIENCE AT THE FRONTIER

1

The Great Heat Engine: Modeling Earth's Dynamics

The Earth embodies the paradox at the heart of geology: a seemingly solid, relatively fixed subject for study, it is actually a dynamic planet whose seething, continuous undercurrent of activity is manifest in the earthquakes and volcanoes that science has only begun to be able to model and predict. These and other cataclysmic symptoms of the Earth's internal dynamics command attention, in part, because they transpire within human time frames—minutes, hours, and days. They make news. Geologic time, by contrast, is measured in millions and billions of years and yet marks the unfolding of an even more dramatic scenario—the movement of entire land masses over thousands of kilometers, though at a rate of only a few centimeters a year, and the creation of mountains, ridges, valleys, rivers, and oceans as a consequence. Further, since most of the Earth's internal processes develop hundreds and thousands of kilometers below the surface, scientists have only indirect evidence on which to base their theories about how these many phenomena tie together.

The search is on for a harmonious model of geologic evolution that will explain and embrace all of the Earth's internal features and behavior (largely inferred) and its tectonics—the deformations of its crust that form the Earth's surficial features. And while no such comprehensive synthesis is in sight, a number of illuminating models accounting for many of the observable phenomena have been developed in the last 30 years, including models explaining the thermodynamics and magnetohydrodynamics of the Earth's core. A number of the other models and theories of how the Earth works are constructed atop the reigning paradigm of geology, plate tectonics, which has only been widely accepted for less than three decades. Plate tectonics has grown from an intuitive suspicion (when maps produced by explorers showed the uncanny complementary shape of the continental shores on either side of the Atlantic, as if the continents had been unzipped), to an early-20th-century theory called continental drift, to the major set of ideas that unify today's earth sciences.

Plate tectonics is no causal force, but merely the surface manifestation of the dynamics of this "great heat engine, the Earth," said presenter David Stevenson. These surface manifestations concern civilization in an immediate way, however: earthquakes, volcanoes, and unpredicted disasters continue to cause tragic loss of life and property.

Stevenson, chairman of the Division of Geological and Planetary Sciences at the California Institute of Technology, described the scientific unifying power of plate tectonics and how it has transformed geology into a brace of strongly interdisciplinary approaches: "Geologists look at morphology, at the surface of the Earth, to try to understand past movements. Petrologists examine the rocks found at the surface of the Earth, trying to understand the conditions under which that material formed and where it came from. Volcanologists try to understand volcanos. Geochemists," he continued, anticipating the discussion of recent diamond-anvil cell experiments, "look at traces of elements transported upward by the forces within the Earth, trying to discern from them the history of the planet, its dynamic properties, and the circulation within. Seismologists look at the travel times of sound and shear waves for variations in signature that provide clues to both lateral and radial structure within the Earth," one of the primary sources of geological information on the deep planetary interior. "Geodesists look at the motions of the Earth as a planetary body, constructing comparisons with distant reference frames, for example, by radioastrometric techniques, and the list goes on and on," Stevenson added. Most of these specialists feel the crucial tug of one another's insights and theories, since the Earth as a coherent system of features and phenomena may be best explained with reference to a variety of probes and explorations.

The Frontiers of Science symposium was host to a solid contingent of such earth scientists, many of whom have been major contributors to the lively debate surrounding the newer theories arising from plate tectonics. Stevenson's presentation, "How the Earth Works: Techniques for Understanding the Dynamics and Structure of Planets," served as a broad overview of the field. Emphasizing the lively premise that "ignorance is more interesting than knowledge," he returned often to the theme of describing Earth processes that "we don't understand."

Marcia McNutt, a session participant from the Massachusetts Institute of Technology, has led numerous marine geophysical expeditions to the South Pacific and has contributed significant new additions to the theory of hot spots. Hot spots, upwelling plumes of hot rock and magma that emanate from the deeper layers of the Earth, may provide important clues to one of the basic features of the Earth's dynamics, mantle convection, a process modeled by another session participant, Michael Gurnis from the Ann Arbor campus of the University of Michigan. Another contributor, Jeremy Bloxham from Harvard University, has conducted an exhaustive study of the Earth's magnetic field by reexamining worldwide historical data spanning centuries and has made some strong inferences about the Earth's core. This deep region has also been explored by simulating the conditions of high pressure and temperature that prevail there, an approach described for the symposium's audience by Russell Hemley of the Carnegie Institution of Washington, where pioneering work has been accomplished in the Geophysical Laboratory through the use of diamond-anvil cells.

The session was organized by Raymond Jeanloz of the University of California, Berkeley, and Sue Kieffer from Arizona State University, whose departmental colleague Simon Peacock was also present. The presenters were also joined in the audience by a pair of prominent geologists from the University of California, Santa Cruz—Elise Knittle, a collaborator with Jeanloz, and Thorne Lay, who has provided pioneering seismological insights about the boundary between the Earth's central core and the mantle above it, a region often referred in its own right as the core-mantle boundary. Together these geologists from many allied disciplines ranged over many of the issues in modern earth science, touching on such basic questions as how the Earth came to be, what it looks like inside, how it works, what early scientists thought about it, and how such views have evolved, as well as what theories now dominate scientific thinking and how scientists develop and test them. In considering these and other questions, they provided a vivid picture of a science almost newly born, one in which the technological and conceptual breakthroughs of the latter half of the 20th century have galvanized a truly interdisciplinary movement.

HOW THE EARTH WORKS

The universe was created at the Big Bang, probably about 13 billion years ago, but more than 8 billion years passed before the cosmic junk cloud where we now reside—itself a residue of the explosive formation of our Sun—began to coalesce into what has become a dynamic and coherent planet, the Earth. As ever larger pieces of the debris called planetesimals aggregated, their mutual gravity and the orbital forces of their journey around the Sun became more attracting, and the Earth grew to what would prove to be a crucial size. During this process, meteorites of all sizes crashed into and became a part of the surface, and much of the energy of such collisions was converted to heat and was retained in the growing mass, or heat sink. Beyond a certain point where size, mass, and heat reached critical dimensions, the collected heat began to generate an internal dynamics, one that continues to this day. The Earth's internal temperature reached a point comparable to that of the Sun's outer regions, and a central core developed that 4.6 billion years later is still about 20 percent hotter than the Sun's surface.

This central furnace probably melted everything, and the iron then sank, relative to lighter material such as silicates, which rose toward the surface, hardened, and became rock. This intense heat energy continues coursing outward through the 6370-kilometer radius of the planet. The Earth also has a second source of energy, the decay of radioactive materials deep within. This atomic process is also converted to heat, and most geophysicists believe this the greater of the two sources of energy powering the heat engine. Regardless of its source, however, it is the *behavior* of this heat that determines the dynamic fate and future of the planet.

A New Model of the Earth

Stevenson surveyed a broad picture from his position in the earth sciences community and provided a cutaway snapshot of the Earth as scientists now believe it to be (Figure 1.1). Although slightly oblate, the Earth may be viewed as a large sphere consisting of more or less concentric layers (although deformities of this regularity exist at the boundaries between the major regions and are of especial interest). The heat engine aspect of the Earth corresponds to this diagram: the closer to the center of the Earth's core, the hotter the temperature and the greater the pressure, reaching a peak of perhaps in excess of 6600°C and over 3.65 million times the atmospheric pressure found at the surface. Thus, at any specific depth, local environmental conditions facilitate and to a limiting extent determine the presence, phase state, and "movability" (or its inverse, viscosity) of particular components of the Earth's interior. Geologists taking seismological readings are most often deducing temperature and pres-

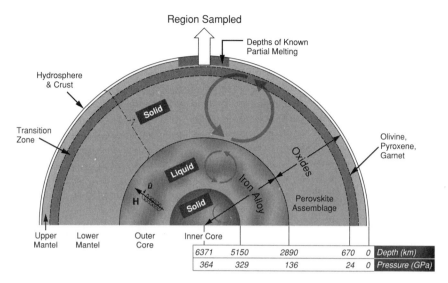

FIGURE 1.1   Schematic cross section of planet Earth. (Adapted from Jean-loz, 1990.)

sure conditions at particular locations within the Earth. However, since all such markers are inferred rather than measured, kilometer readings of depth render a more graphic image. Stevenson reinforced the importance of the underlying regular phenomena of temperature and pressure. Together with measurements of the Earth's magnetic field, they provide baseline information for modeling that allows scientists to search for anomalies.

At the center of the Earth, 6370 kilometers from its surface, is its inner core. "The solid inner core has seismic properties consistent with those of metallic iron," said Stevenson, who added that "one suspects there is some nickel mixed in." Although it is the hottest region within the Earth, the inner core is solid due to the astounding pressures it is subjected to. Upward from the center toward its 2900-kilometer-deep outer edge, the core changes to a liquid. Slightly less pressure and—some believe—possibly higher proportions of other materials permit this change. This liquid is in motion, due to the heat convecting from beneath, and may be said to be sloshing around a million times more rapidly than material in the inner core beneath it. The motion of this material, said Stevenson, "is the seat of the dynamo action, where the Earth's magnetic field is formed." Thus more than half of the Earth's diameter consists of the inner and outer core. Next comes a very small region rich with phenomena and controversy, the core-mantle boundary (CMB) between the core and the next major section above, the mantle. A number of regional structures of as much as several hundred kilometers seem to arise discontinuously near this depth and are referred to as the D-double prime (D") region. The temperature here may be perhaps 3500°C, but is quite uncertain.

Above the CMB (and—where it is found—the D") is the other major region of the Earth's interior, the mantle, which rises from a depth of 2900 kilometers through a transition zone at about 670 kilometers to its upper edge, variably between depths of about 50 to 150 kilometers. Thus the core and the mantle together make up over 99 percent of the Earth's volume. "The mantle is mostly solid, composed of materials that are primarily magnesium, silicon, and oxygen—the silicates and oxides," said Stevenson. The final of the Earth's three layers is the crust, whose thickness ranges from about 6 kilometers under the oceans to 50 kilometers under certain continental regions. Dividing the Earth into core, mantle, and crust is traditional, and these three regions are distinct in their rock chemistry.

Alternatively, geologists tend to divide the outermost part of the Earth into regions that reflect differences in the ways materials behave and how matter deforms and flows—the study of rheology. The uppermost region, the lithosphere, is rigid, is on average about 150 kilometers thick (thereby embracing the crust and part of the upper mantle), and actually slides around on the surface of the top zone of the upper mantle, known as the asthenosphere, which convects and is considered to be less viscous than the zones immediately above and below it, and may be partially melted in some places. Generally about 200 kilometers thick, the asthenosphere is the earthly "sea" on which float the lithospheric plates, atop which civilization and the Earth's oceans and its visible surface deformations perch. The lithosphere is not a solid, continuous shell but consists rather of at least a dozen identifiable plates (and more, when finer distinctions are made) that quite literally slide over the surface of the asthenosphere.

How the Model Came to Be

People have no doubt always wondered about the Earth on which they stood, but the science of geology was born from the observations and ideas of European and North American naturalists near the turn of the 19th century. When British geologist Charles Lyell (1797–1875) in his classic *Principles of Geology* developed what he called uniformitarianism, he struck a new course from the prevalent view that the Earth was formed by cataclysmic events like Noah's flood, and other biblical occurrences. He suggested that fundamental ongoing geologic processes were the cause of the Earth's primary features—its mountains, valleys, rivers, and seas—processes that he believed had been occurring gradually and for much longer than was then thought to be the age of the Earth. The prevalent view that followed for nearly a century and a half, however, was that the Earth's surface was fairly rigid.

The notion that the Earth's solid surface might be mobile—"although on geologic time scales of hundreds of millions of years," said McNutt—had been a distinctly minority opinion, first proposed by biologists like Charles Darwin who could in no other way account for similarities of flora and fauna at widely separated locales. Then another science, physics, began to chip away at the theory of a rigid Earth. The development of radioactive dating methods provided a way of discerning the age of a material by its rate of atomic decay, and suddenly the assumption made by Kelvin and others that the Earth was at most 100 million years old was obliterated, and along with it mathematical calculations of the cooling rate. Rocks could now be dated and showed age variations with consistent pattern over hundreds of millions of years. Suddenly a whole new set of data was presented to theorists, and geologists now had a time line for the Earth's history that was consistent with mobilism, the theory of dramatic—albeit slow—movements within the Earth over long spans of time.

Plate tectonics, like many another revolutionary idea, has had a long genesis. As did their colleagues in biology, late-19th-century geologists found striking similarities in structure and materials throughout the Southern Hemisphere and theorized about an erstwhile supercontinent, which Austrian geologist Eduard Suess named Gondwanaland. Looking at all of these clues, German geologist Alfred Wegener (1880–1930) proposed a formal theory of continental drift in 1915. Although many observed, based on evidence provided by ever better maps, that the outlines of the continents seemed to dovetail—as if they were separate pieces broken off from one original continuous land mass Wegner called Pangaea—Wegener's ideas were not embraced for decades, and he died on expedition in Greenland, in search of corroborating evidence that that island was indeed drifting away from Europe.

By the 1960s, observations of a great Mid-Atlantic ridge on the ocean floor almost precisely midway between Europe and America revealed it to be a crack in the Earth, from which spewed molten rock. A magnetic profile of the material away from the crack toward the respective continents shows a steadily older rock, indicating that the plates on which the continents rest are moving apart at a rate of about 2 centimeters per year. Scientists now reason with much greater confidence that 130 million years ago the two continents were indeed one, before a series of volcanoes and earthquakes developed into a rift that filled with water from the one primordial sea, giving birth to distinct Atlantic and Pacific oceans. The ridge has since been recognized as the longest structure on the Earth, over 75,000 kilometers, winding from the Arctic Ocean through the Atlantic, eastward around Africa, Asia, and Australia, and up the Pacific along the West Coast of North America. "Actually a system of mid-ocean ridges," said Stevenson, it provides geologists with a major geological feature that McNutt believes deserves more exploration.

The Earth's dozen major lithospheric plates illustrate demonstrably at their boundaries their underlying movement with respect to one another (Figure 1.2). At divergent boundaries such as the Mid-Atlantic ridge, as the plates move apart in more or less opposite vectors, molten material erupts in the rift and forms a mountainous ridge, actually adding to the Earth's crust. Conversely, at convergent

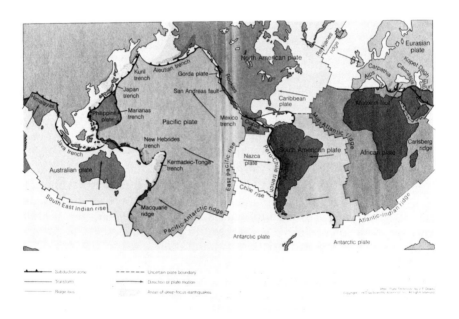

FIGURE 1.2 Distribution of the Earth's major surface plates. (Reprinted with permission from Press and Siever, 1978, after "Plate Tectonics" by J.F. Dewey. Copyright © 1972 by Scientific American, Inc.)

boundaries such as the Marianas Trench, the plates move directly toward one another, and as they collide, one slides directly above the other, the latter undergoes subduction, and crustal material is destroyed by being recycled into the mantle. If the vector directions of the plates are neither in collision nor opposition, the plates may be said to slide past one another in what is called a transform fault. The Earth's total mass and its surface area are conserved in the plate tectonic process.

Rather than describing it as a full-blown theory, Stevenson preferred to label plate tectonics as "a description of what happens at the Earth's surface—the motion associated with continental drift and the generation of new ocean floor. It is a phenomenological description." He did not diminish its power, however. Clearly, he said, "the paradigm governing much of earth science since the 1960s is mobility. It is a picture particularly pertinent to an understanding of the ocean basins. A variety of measurements, initially paleomagnetic and subsequently geodetic, have confirmed the picture rather well."

NEW METHODS AND TOOLS FOR WORKING GEOLOGISTS

The rapid progress in modeling of the Earth in the last 15 years owes a heavy debt to growing computer power, especially in the analysis of seismic waves, which have been collected and studied without the aid of sophisticated computers since the 1920s. Closely related to this improvement have been advances in the data collection itself, with more—and more accurate—measuring instruments and seismic wave collection centers being established all over the world, and a better and more elaborate campaign of magnetic and gravitational data surveys as well. Moreover, simulations are not limited to the realm of the computer: a new process called diamond-anvil cell technology has been developed to recreate the high-temperature and high-pressure conditions thought to exist in the lower mantle and the core itself.

As suggested by Stevenson's catalog of allied disciplines, geology employs a number of distinct scientific tools and methods to model and probe the Earth's interior. One of the most valuable, barely a decade old, is seismic tomography, which exploits the sound and shear waves created each year by many global earthquakes of varying intensities. An earthquake, although difficult to predict, is not hard to rationalize. Again, it is the Earth's dynamism, specifically its property of imperfect elasticity, that gives rise to seismic waves. An elastic medium will resist deformation, such as that promoted by the moving plates. This resistance to being compressed or sheared takes the form of a restoring force, which, when the break finally occurs because of imperfect elasticity, is released to emanate through the medium in the form of a seismic wave, i.e., a strong, low-frequency sound wave. These waves take several different forms, but each has a recognizable signature and behaves with distinguishing characteristics according to the principles of physics. Surface waves, called Love and Rayleigh waves, travel circular paths along the Earth's surface and extend deep enough into the Earth to interact with the upper mantle. So-called body waves forge curved paths through the Earth's depth and deliver information from their journey all the way down to the core.

Seismic tomography employs heavy computer power in a process closely analogous to computerized tomography in medicine. Instead of undergoing variable absorption, as do x rays in the body, seismic waves confer information by altering their speed when they encounter differing materials. As with computerized tomography, the results provide a three-dimensional picture of the waves' journey from their source at the epicenter of a quake to the many measuring devices spread all over the globe. Since the journey's distance can be computed precisely, any delay in a wave's arrival at a given site can be attributed to the nature and condition of the media through which it just passed. A given wave may penetrate thousands of kilometers of the Earth on its journey and arrive with but one piece of data, its travel time, which yields only an average for the trip, not data indicating how the wave may have slowed down and speeded up as it encountered different conditions along its path. Thousands of such readings all over the planet for similar waves traveling along different, cross-cutting paths make it possible to perform extremely sophisticated analyses of the data, since each separate measurement can provide a restraint on the generality of each of the others. Elaborate mathematical analyses combine all of these data into hitherto unglimpsed, three-dimensional views of the inner Earth, all the way down to the center.

Seismic tomography exploits the fact that a medium's physical properties—its density, composition, mineral structure, mobility, and the degree of melt it contains—will determine how fast waves of a particular type are transmitted through it. The liquid material in the Earth's outer core does not transmit shear waves but does transmit compressional waves. Properties such as a medium's rigidity and its compressibility are in turn affected by its temperature and its pressure. Knowing the type of wave being measured and the distance it travels, seismologists can hypothesize based on what they believe they know about temperature, pressure, mantle convection, and the composition of the Earth at different depths. When the tomographic readings vary from the prediction, anomalies can be inferred and theories developed to explain them that nonetheless fit all of the constraining data. Colder material tends to be stiffer and a faster transmitter of waves than hotter material.

Another property providing seismologists with valuable data is how a material's mineral crystals are aligned. Flowing material in the mantle tends to orient the crystals in nearby rock in the direction of the flow. Crystals have three different axes that affect the speed of seismic wave propagation differently. If randomly aligned, the crystals produce an average wave transmission speed, compared to a faster speed if a crystal's fast axis is parallel to a given wave's (known) direction of propagation. The complexity of this picture does not yield perfectly transparent readings, but the power of the computer to filter the data through a web of interrelated equations provides seismologists with a powerful modeling tool.

Geological models of the deep Earth must incorporate the lessons of solid-state physics, said Stevenson, which indicate that most materials will behave differently under the temperature and pressure extremes thought to prevail there. Some of these assumptions can be tested with a new technology that promises to open fertile ground for theorizing. A diamond-anvil cell consists of two carefully cut and polished diamonds mounted with their faces opposite each other in a precision instrument capable of compressing them together to produce extremely high pressures, creating an "anvil" with a surface less than a fraction of a millimeter across on which samples can be squeezed (Figure 1.3). Hemley, who uses a variety of such devices at the Carnegie Institution's Geophysical Laboratory in Washington, D.C., explained that the transparent nature of diamonds allows the experimenter to further probe the sample with intense laser and x-ray beams. Lasers can also be used to increase the temperature. Together, these techniques can simulate temperatures approaching those found at the surface of the Sun and pressures over 3 million times that at the Earth's surface, up to and including the pressures at the center of the Earth.

"Under these conditions, planetary constituents undergo major structural and electronic changes," explained Stevenson. The Earth's most common mineral is probably magnesium silicate perovskite, a denser structure type than most other constituents and one that is not thermodynamically stable for silicates at normal atmospheric pressure. The normal olivine structure—in which a silicon atom is surrounded by four oxygen atoms—has been shown by diamond-anvil cell experiments to undergo a series of transformations, ultimately

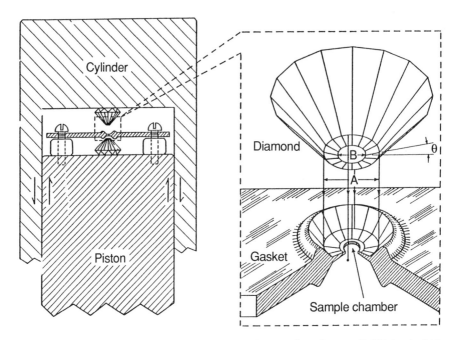

FIGURE 1.3 Diamond-anvil cell with multimegabar design (7 GPa). A, 300 to 500 μm; B, 25 to 250 μm; and Θ, 4 to 10°. (Courtesy of R. Hemley.)

taking on the perovskite structure, in which the silicon is surrounded by two additional oxygen atoms. This occurs at just about the pressures that have been determined by seismology to prevail at the 670-kilometer boundary region between the upper and lower mantle (Knittle and Jeanloz, 1987). Said Stevenson, "The reduction of bandgaps and preference for compact phases may greatly change our view of the differences between and interrelationships among silicates, oxides, and metallic oxides."

Hemley believes that discovery of these phase transitions and measurements of properties of the perovskite minerals favor the mantle convection models that employ more complicated patterns than the single, deep-convection theory. "But," he cautioned, "more measurements are required to be certain." Hemley's laboratory and others continue to probe the atomic behavior of materials under conditions more like the extreme ones found deep inside the Earth. One such material is hydrogen, interest in which, said Hemley, "dates back to the 1935 calculation by Wigner and Huntington that showed that a monotonic lattice of hydrogen atoms will be a high-density metal. Hundreds, if not thousands, of papers have been written on this topic in the intervening 55 years." Hemley explained that until recently, direct tests were not possible, and so he and his colleague H.K. Mao wanted to test their prediction about where "the transition from the low-density, molecular, insulating phase to this metallic phase" would occur.

Although exploring the metallic phase of hydrogen might seem to be a concern primarily of condensed-matter physics, this question has great relevance to planetary geology as well and illustrates, said Hemley, the interdisciplinary nature of this work. Jupiter and Saturn are composed largely of hydrogen and are believed to contain metallic hydrogens because of the enormous pressures reached inside. "Understanding the metallization phenomenon is therefore crucial for modeling their dynamics," explained Hemley. So far, their results indicate an intriguing series of phase transitions rather than simple conversion of an insulating molecular solid to an atomic metal. These new phenomena are challenging them to invent new variations on the basic diamond cell techniques and are driving the development of new theories in condensed-matter and planetary sciences.

Gravity provides another lens into the Earth. Since Newton's time, the understanding of gravitational forces has provided basic constraints as well as a realm of useful data for scientists in many fields, for geologists perhaps more than others. Scientists refer to the geoid as the equipotential surface normal to the Earth's gravity. "A buried mass excess," explained McNutt, "will raise the geopotential surface, that is, cause it to move further from the Earth's center. Conversely, a mass deficit will lower the equipotential."

"Geodesy remains a powerful method," said Jeanloz, and together with seismology offers a powerful way of looking inside the Earth, notwithstanding the recent applications of tomography. All of these methods provide excellent and often distinct supplementary perspectives to corroborate or constrain theories based on data gathered with yet other methods. In a science for which the deepest mine penetrates barely 2 miles and the deepest hole less than 10, inferences assume greater importance because most measurements are necessarily indirect. And while geophysics is providing richer data with each year of technological progress, geologists searching for the big picture must bring together a wealth of distinct, sometimes contradictory perspectives. Theorizing about how the Earth works is inviting and yet risky. But the ferment and excitement surrounding the endeavor are palpable, as the Frontiers session on geology illustrated.

SHARPENING THE FOCUS ON THE EARTH'S DYNAMICS

A central problem for earth scientists, said Stevenson, is to reconcile plate tectonics and mantle convection. Both phenomena occur, but exactly how convective currents inside the mantle may produce the motions of the plates has not been fully clarified. Because the Earth's internal heat source has been generally verified, he continued, our knowledge of the physical principles of convection makes this explanation appealing on the theoretic level. Undoubtedly the mantle convects; the question is what patterns the convection follows. The question is by no means trivial, for if there is one large convective pattern circling throughout the mantle—the deep-convection model—the fate of the Earth and its internal heat will run differently than if there are at least two layers more or less independently convecting. On the other hand, while plate tectonics has been demonstrated by myriad data, measurements, observations, and chemical analysis, a full theoretical explanation has yet to be offered. Geologists have not been able to come up with a comprehensive model that incorporates either mantle hypothesis with plate tectonics.

As Stevenson put it, "The biggest problem confronting a synthesis of plate tectonics and mantle convection lies in the inability to incorporate the extremes of rheology," the study of the deformation and movement of matter. A problematic element for the mantle convection models is the quasi-rigid nature of the plates. Those models posit that temperature has a great effect on viscosity, but that would lead, he said, "to a planet completely covered with a surficial plate that has no dynamics at all," despite the convection beneath the surface, instead of the planet we actually observe. Thus the appealing theoretical success of mantle convection does not begin to explain much of the data. Stevenson called it "a profound puzzle," and concluded, "We simply do not understand why the Earth has plate tectonics."

Yet the plate tectonics model also embodies a dilemma: not only is the cause of plate tectonics elusive, but conventional models also do not even fully describe the phenomenon. As Stevenson said: "An interesting question is the extent to which plate tectonics is even correct. That is to say, are those plates actually rigid entities rotating on the surface of the Earth about some pole? The answer," he continued, "is yes, to perhaps a 90 percent accuracy. You might say that that is quite good, but that 10 percent failure is important because it leads to the formation of many of the Earth's mountain ranges and other prominent geological features." Scientists modeling plate tectonics treat the plates as rigid and develop predictions about how they move over the Earth's surface. The 10 percent failure to predict what is observed points to "deformation primarily at plate boundaries," said Stevenson, and explains why so much attention is paid to regions of the world where the plates are colliding. Another phenomenon that may provide a link between convective and plate tectonic theories is the world's 40 or so active hot spots.

Hot Spots, French Polynesia, and the "Superswell"

Hot spots present scientists with what seems to be another main ingredient in the Earth's recipe for its simmering equilibrium. These plumes emanate from deep within the mantle (or some from not so deep) and consist of slender columns, each about 300 kilometers in diameter, of slowly rising rock hotter than their surroundings (Figure 1.4). Although the temperature differential may be as small as 100°C, its impact on viscosity and therefore on the rise of the convective plume can be dramatic. Most hot spots are believed to begin as deep-mantle phenomena and provide a benchmark for plate movement because of their small lateral motion. Since the lithospheric plates move across and above the asthenospheric point(s) where a plume emerges from its long journey through the mantle, they get marked by a trail of volcanoes that leave a permanent record of the direction and speed of the plate itself.

The first coherent theory of hot spots was proposed by the University of Toronto's J. Tuzo Wilson in 1963 after he studied the Hawaiian volcanic fields. In the intervening years, the plume model has been elaborated and now may provide an explanation of how convection delivers heat to the surface, although its most recent proponents hasten to admit that no plumes have been observed directly. Vink et al. (1985) have explained that "ultimately the energy that

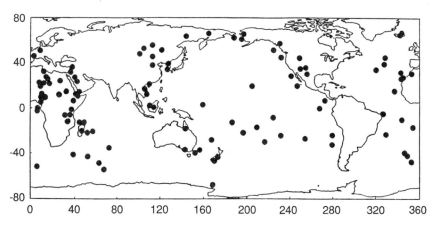

FIGURE 1.4 Distribution of surface hot spots, centers of intraplate volcanism, and anomalous plate margin volcanism.

drives plate motion is the heat released . . . deep in the mantle. The plumes provide an efficient way of channeling the heat toward the surface. . . . Less viscous material produced by variations in temperature or volatile content tends to collect and rise toward the surface through a few narrow conduits, much as oil in an underground reservoir rises through a few bore-holes" (p. 50–51). Their theory does not contend that the plumes actually propel the plates, but rather that "the two are part of the same convective cycle" (p. 51). The asthenosphere must be heated in order to provide a less viscous surface for the plates to slide across, and without the plumes to provide this heat the plates would grind to a halt.

Certain regions of the Earth's surface are pregnant with these hot-spot eruptions from the mantle beneath. One such region is the seafloor beneath French Polynesia in the South Pacific, explained McNutt, whose career has been devoted to looking more closely at the hot spot phenomenon there. That region differs from the standard oceanic lithospheric profile "in almost all respects. Its depth is too shallow, its flexural strength too weak, the velocity of seismic waves through it too slow, and geoid anomaly above it too negative for what has been determined to be its lithospheric age," she pointed out. Together with MIT co-worker Karen M. Fischer, McNutt named this geologic curiosity the South Pacific "Superswell."

Theorists wonder if this area might yield clues to a unified theory. By positing a thinner plate in this region, McNutt can account quantitatively for the anomalous seafloor depths observed and qualitatively for the low velocities of seismic surface waves, the weak elastic strength of the lithosphere, and the region's high vulnerability to what is distinguished as hot-spot volcanism. The thinner plate hypothesis does not, however, account for other observed features, such as why the plates move faster and spread faster over the mantle in the Superswell region than anywhere else on the Earth. "And most importantly," emphasized McNutt, "a thinner plate in the absence of other explanatory features would produce a geoid high," not the low that is actually observed.

Polynesia presents another mystery to earth scientists. Seismic studies confirm that these volcanoes are of the hot-spot variety. Although constituting only 3 percent of the planet's surface, the region accounts for 30 percent of the hot-spot material—the flux of magma—that appears all over the Earth. But why there, and why now? The answer, or at least another important piece of the puzzle, may lie thousands of miles to the northwest in an area of ocean floor southeast of Japan known as the Darwin Rise. More hot-spot volcanoes are clustered there, although these are now extinct and lie beneath the surface of the sea. What is compelling about these two fields of volcanism—one active, the other extinct—is their location. The Darwin Rise tests out at an age of 100 million years. McNutt was able to "backtrack the movement of the Pacific plate in the hot-spot reference frame to its position at 100 million years ago," and she demonstrated that when the hot-spot volcanoes of the Darwin Rise were created—when material welling up from within the Earth formed a ridge with active volcanoes—it was located above the mantle where the French Polynesian volcanoes are now erupting. She argued from these data that hot spots must be driven by processes within that part of the mantle and, further, that they occur episodically.

To account for these observations, McNutt and co-workers at MIT have developed a theory. Seismic wave data show that the lithosphere beneath the Superswell has been thinned to about 75 kilometers, 40 percent less than the usual lithospheric thickness under the North Atlantic and North Pacific oceans. This thinner plate, she believes, is not the cause of the Superswell; rather it is the surficial manifestation of forces below, specifically a low-viscosity area under the plate and a plume of heat convecting through the mantle. To simplify somewhat, the MIT team has inferred from the data that a large hot blob of material periodically rises through the mantle from near the CMB in the Superswell region. Once the blob arrives near the upper reaches of the mantle, one would expect the geoid above to reflect it. Since the data show a depression—not a rise—in the geoid, something else must be happening. McNutt has inferred the presence of a low-viscosity zone near the upper reaches of the mantle. Such a phenomenon would account for the geoid anomaly and further might explain why the lithosphere in the Superswell region moves so rapidly.

Is Volcanism an Essential Part of the Cycle?

Hot-spot volcanoes may well be an indigenous phenomenon, pointing to something unique in the Earth below them. But volcanoes erupt all over the Earth, and Stevenson believes it important to understand the role volcanism plays in the interaction between the tectonic plates and the mantle. He said that thermal convection theories alone are not rich enough to explain all the complexities of the Earth's dynamics. "It seems unlikely that we can claim any deep understanding of how the Earth works if we confine ourselves to the behavior of a single working fluid of very high viscosity, and restrict ourselves to density differences that arise only from the small effects of thermal expansion," he explained. While these considerations are not fully accepted tenets in earth science, they have deflected inquiry away from another phenomenon Stevenson thinks suggestive: differentiation of the Earth's constituents and thus the possibility of two-phase flow. In theory, convection could be occurring up through the Earth's profile with no actual melting into liquid form. It is known that partial melting is consistent with the fact that some of the components melt at lower temperatures than do the dominant constituent. From such melting—at least in localized regions—result significant changes in fractional density, up to as much as 10 percent.

Now the deductions start to line up: changes in density from melting or partial melting produce significant changes in viscosity; the melting process can be enhanced by water or other volatiles; the result, said Stevenson, echoing the plume model, may be a kind of "lubrication of plate tectonics." That is, the presence of partially melted lower-viscosity material at the boundary may effectively "decouple the mantle from overlying plates."

So, according to this hypothesis, partial melting helps lubricate the sliding plates, which at moments and points of impact may spew volcanoes. What, then, would happen if there were insufficient heat in the Earth to accomplish this melting? If the temperature of the mantle were reduced by only 5 percent, said Stevenson, volcanism would largely cease: "If that occurred, what then would the Earth do? The Earth is a heat engine. It could certainly keep convecting, but could it have plate tectonics?" He reported that many of his colleagues suspect "that the melting process, the formation of basalt, may be an important factor keeping the plate tectonics system going and may be a part of the difference between a planet that has a single plate completely encompassing it and a planet, like the Earth, which is tessellated into these many plates of many different sizes." For Stevenson, therefore, volcanism is essential: "It is not a sideshow. To understand mantle convection, you have to take into account the fact that material melts to some extent."

Dynamics in the Core and the Magnetic Field

Looking deeper yet, beneath the mantle, scientists find that the material in the outer core is highly conductive and low in viscosity. The forces of convection moving through this region have somehow set up a continuously regenerating dynamo, a massive natural version of the man-made power systems up above that produce predictable magnetic fields. Exactly how this dynamo works is not known, although a semiquantitative picture has developed over the past few decades. Stevenson pointed out that scientists attempting to understand the Earth's magnetic field have one advantage over those trying to understand the interaction between the mantle and tectonic plates: the material parameters of the core may be simpler and less variable. Nevertheless, Stevenson emphasized, "This is a difficult problem. People are tackling it now on supercomputers and by approximation techniques."

The magnetic field profile of the Earth shows a familiar dipole. The geomagnetic field actually observed, however, contains a number of other, more complex components. Often in geology scientists look to the edges, the boundaries, for their cues, and it seems likely that some of the magnetic field anomalies are related to what is happening at the CMB. "There must be a bumpy and fuzzy interface between the core and the mantle," Stevenson said, rather than "the fluid dynamicist's ideal world of a smooth, isothermal, equipotential yet rigid bounding surface." In fact, the terrain at the bottom of the mantle known as the D"—a region, where it occurs, of a few hundred kilometers with characteristic, perhaps even continental-scale features—could be comparable in general shape to the topography seen at the surface. If this region is truly rich with such features, Stevenson continued, it follows that heat flow between various regions of the mantle and the core would be nonuniform. And the effects would not stop there.

These features at the interface most likely have a significant influence on heat flow in the outer core, perhaps on the heat flow that generates the dynamo itself. This raises the exciting but complex prospect of relating the observed field geometry and how it has varied throughout a couple of hundred years of recorded history to the distribution of irregularities in the lowermost mantle. For example, take a region of mantle that has substantial downflow because it is colder than average. What can be inferred about the CMB in this local region? Stevenson believes that the core probably contains "a very-low-viscosity fluid that may vigorously convect on a small-length scale . . . in the same way as air shimmers in the desert in the boundary layer region between the hotter Earth and the cooler atmosphere." This border region draws contending theories as honey does bees, but few as controversial as that of Jeanloz.

Jeanloz's work in probing the chemical reactions at the CMB has been provocative but has enriched the debate considerably. Along with Knittle, Jeanloz believes that there is considerable chemical interaction between the silicates that probably make up the lower mantle and the iron of the outer core. As Hemley and the diamond-anvil cell experiments have shown, these elements remain chemically aloof at the surface, but under conditions of extreme pressure and temperature they intermix. Thus, Jeanloz and Knittle have reasoned, silicon or oxygen alloys from the lower mantle could actually become a part of the core's chemistry and thereby reduce its density (Knittle and Jeanloz, 1991), a finding corroborated by other independent measurements. Thorne Lay has provided a powerful set of seismic measurements that seem to corroborate material variations.

Stevenson cited recent analyses of data in ships' logs and marine almanacs suggesting that for almost 300 years the basic features of the geomagnetic field have remained fixed relative to the mantle, "indicating at least partial control of the field by mantle irregularities." That work has been conducted by another session participant, Jeremy Bloxham of the Department of Earth Sciences at Harvard University. Rather than taking a top-down, mantle-oriented approach, Bloxham has looked at the same complicated magnetic field inferred to exist at the CMB for implications about the dynamics of the deeper core. Satellite measurements of the geomagnetic field taken in 1969 and again in 1980 showed small changes. Plugging these data into the equations that describe the Earth's magnetic field allows scientists to construct a map of the fluid flow just beneath the surface of the core. Bloxham hastened to warn, however, what a small pedestal these observations provide for scientists to try to view the Earth's history, since the velocities of the liquid iron that generates these fields are on the order of 20 kilometers per year (10^{-3} m/s): "If we draw an analogy with meteorology where wind speeds in the atmosphere are on the order of meters per second, then—in a very crude sense—we can say that looking at the atmosphere for a day yields the same insight as looking at the core for 60 years. In other words, we need 60 years worth of observations of the core to get the equivalent information that we could obtain observing the atmosphere for a single day." Conversely, theories based on the patterns of flow in the core derived from this 11-year data slice might be compared, Bloxham said, to "trying to understand the dynamics of the atmosphere based on just a few hours of data."

The good news is that flow patterns are observed to change only slowly as one penetrates deeper into the core. Thus the picture drawn from the satellite data may well correspond to the full profile of the outer core, said Bloxham, who has also found corroboration from a surprising source. Field researchers for the past 300 years have been making reliable measurements, and some data obtained as long ago as the late 17th century can now be reinterpreted through modern theory. Bloxham said that these data suggest several fairly constant features of the geomagnetic field. For one, the field at the CMB beneath the North Pole is consistently close to zero, a feature inconsistent with a simple dipole model of the magnetism in which the field should produce maximum radial flux at the poles. Also, certain concentrations of high flux at the Earth's high latitudes have remained fairly constant over time.

Theory and experiment indicate that a rapidly rotating fluid forms convection columns. These columns align parallel to the axis of rotation. Bloxham said that one interpretation of the magnetic field data suggests that such a process is occurring in the outer core, with large convection rolls aligned parallel to the Earth's rotation axis then grazing the solid inner core. The observed, high-latitude, surface concentrations of magnetic flux would, according to this interpretation, correspond to the tops and bottoms of these convection rolls, while the regions of almost zero flux at the poles also fit the pattern predicted from such an effect. Bloxham has become one of the authorities on the core, and his view is that no significant layering occurs there.

Evidence in the Fossil Record

Earth scientists face another dilemma. Despite dramatic advances in the arsenal of scientific measuring techniques, data on the Earth's dynamics can only measure what is happening today. Yet the Earth has been evolving for billions of years. As session participant Michael Gurnis put it, "Current measures of the planet, as powerful as they are, are weak in terms of understanding the true dynamics of the Earth." Geologists have known for many years that mantle changes can move continents and cause fluctuations in sea level over time. This ebb and flow leaves a pattern in the rocks examined on continents because, when the shore at a continent's margin is submerged, characteristic sediments accumulate and harden into rock. Over long periods of geologic time, such rocks record this history of ocean flooding. Gurnis pointed out that "geologists have recognized for the past century that these rock formations are measuring changes in the Earth's shape," and that the sedimentary rock patterns on the offshore continental shelves provide a four-dimensional map of the movement of the ocean's edge back through time.

Continental drift does not seem to be the sole factor defining the ocean's edge. Continental crust, composed of material like granite, has a lower density than the ocean floor crust, which is rich in basalt, and so it floats higher on the mantle than do other regions. Thus arises a different approach to the earlier question: How does convection in the mantle work together with tectonic plate movement? Theoretical and modeling work on this question suffer, said Gurnis, from our "poor ability to simulate on a computer the interaction between tectonic plates and mantle convection." Assuming simple hydrodynamic convection through a mantle modeled with the appropriate characteristics and physical properties, he and others try to predict what would happen to a single continental plate above. Gurnis said that the model demonstrates patterns of continental flooding that leave a characteristic signature, which then correlates back very well to the internal dynamics of the model system. This effort points to the possible value of the historical record of sea level changes, a potential database on the order of 500 million years old. There may yet come a computer simulation of the planet's interior that will crack the time barrier and write a true geological history of the whole Earth, inside and out.

Geology of the Other Planets

Stevenson began his presentation with the remark that "among the solid planets, the Earth is unusually dynamic." Can this observation be used to further the inquiry into the dynamics themselves? Stevenson's speculations have been fortified by spacecraft providing often dramatic and penetrating views of other planets in our solar system. A series of probes to Mars has revealed many features. The present Magellan mission to Venus is providing a high-resolution radar map of the surface of the Earth's nearest, cloud-shrouded neighbor. And the two Voyager spacecraft have yielded a wealth of data on more distant solar system neighbors—Jupiter, Saturn, Uranus, and Neptune. Bloxham reported that "until 2 or 3 years ago, I think a discussion like this would have included the belief that all magnetic fields we are aware of align with the rotation axes of their planets. This was confirmed with the Sun, Earth, Jupiter, and Saturn." But then "along came information on Uranus and Neptune, which have fields inclined at some 50 degrees to their rotation axes. That threw a small monkey wrench into the works," he pointed out.

But information from the near and far solar system has not yet supplied inferences that solve the mysteries of convection. Scientists know, for example, that all planets, large satellites, and moons convect, but only the Earth among the solid bodies observed appears to undergo the striking characteristics of plate tectonics. Venus might possibly have some plate-tectonic-like features, and the controversy over this question awaits further data from the Magellan mission. Nevertheless, said Stevenson, sliding tectonic plates have been confirmed nowhere else and are not yet fully understood for this planet. "Mars is a one-plate planet," said Stevenson, and "actually even Mercury and the Moon are believed to have mantle convection," although, at present, "it is much less vigorous."

He conceded it is a reflection of our green understanding that despite "vast terrestrial experience" we are not far along the path of fully describing other planets. Far from ripe and seasoned, then, planetology is a new and still profoundly mysterious science. And though he and most geologists may focus more on our own planet, Stevenson believes that developing a generic awareness of the other known planets will enhance and enrich the context of observations made on the Earth.

BIBLIOGRAPHY

Anderson, Don L., and Adam M. Dziewonski. 1984. Seismic tomography. Scientific American 251(October):60-68.

Bloxham, Jeremy, and David Gubbins. 1989. The evolution of the Earth's magnetic field. Scientific American 261(December):68-75.

Glatzmaier, Gary A., Gerald Schubert, and Dave Bercovici. 1990. Chaotic, subduction-like downflows in a spherical model of convection in the Earth's mantle. Nature 347(6290):274-277.

Jeanloz, Raymond. 1990. The nature of the Earth's core. Annual Review of Earth and Planetary Sciences 18:357-386.

Knittle, Elise, and Raymond Jeanloz. 1987. Synthesis and equation of state of (Mg,Fe) SiO_3 perovskite to over 100 GPa. Science 235:668-670.

Knittle, Elise, and Raymond Jeanloz. 1991. Earth's core-mantle boundary: Results of experiments at high pressures and temperatures. Science 251:1438-1443.

McNutt, Marcia K., and Anne V. Judge. 1990. The superswell and mantle dynamics beneath the South Pacific. Science 248:969-975.

McNutt, M.K., E.L. Winterer, W.W. Sager, J.H. Natland, and G. Ito. 1990. The Darwin Rise: A cretaceous superswell? Geophysical Research Letters 17(8):1101-1104.

Powell, Corey S. 1991. Peering inward. Scientific American 264(June):101-111.

Press, Frank, and Raymond Siever. 1978. Earth. Second edition. Freeman, San Francisco.

Scientific American. 1983. The Dynamic Earth. Special issue. Volume 249 (September).

Stein, Ross S., and Robert S. Yeats. 1989. Hidden earthquakes. Scientific American 260(June):48-57.

Vink, Gregory E., W. Jason Morgan, and Peter R. Vogt. 1985. The Earth's hot spots. Scientific American 252(April):50-57.

RECOMMENDED READING

Anderson, D.J. 1989. Theory of the Earth. Oxford, New York.

Fowler, C.M.R. 1990. The Solid Earth. Cambridge University Press, New York.

Hemley, R.J., and R.E. Cohen. 1992. Silicate perovskite. Annual Review of Earth and Planetary Sciences 20:553-593.

Jacobs, J.A. 1992. Deep Interior of the Earth. Chapman & Hall, New York.

Jeanloz, R. 1989. Physical chemistry at ultrahigh pressures and temperatures. Annual Review of Physical Chemistry 40:237-259.

Lay, Thorne, Thomas J. Ahrens, Peter Olson, Joseph Smythe, and David Loper. 1990. Studies of the Earth's deep interior: Goals and trends. Physics Today 63(10):44-52.

2

Artificial Photosynthesis: Chemical and Biological Systems for Converting Light to Electricity and Fuels

"Natural photosynthesis is a process by which light from the sun is converted to chemical energy," began Mark Wrighton in his presentation to the Frontiers symposium. Wrighton directs a laboratory at the Massachusetts Institute of Technology's Chemistry Department, where active research into the development of workable laboratory synthesis of the process is under way. As chemists have known for many decades, the chemical energy he referred to comes from the breakdown of carbon dioxide (CO_2) and water (H_2O), driven by photons of light, and leads to production of carbohydrates that nourish plants and of oxygen (O_2), which is vital to aerobic organisms. What is not known in complete detail is how this remarkable energy-conversion system works on the molecular level. However, recent advances in spectroscopy, crystallography, and molecular genetics have clarified much of the picture, and scientists like Wrighton are actively trying to transform what *is* known about the process into functional, efficient, synthetic systems that will tap the endless supply of energy coming from the sun. "Photosynthesis *works*," said Wrighton, "and on a large scale." This vast natural phenomenon occurring throughout the biosphere and producing an enormous amount of one kind of fuel—food for plants and animals—Wrighton described as "an existence proof that a solar conversion system can produce [a different, though] useful fuel on a scale capable of meeting the needs" of human civilization. Photovoltaic (PV) cells already in use around the world provide a functional (if more costly per kilowatt-hour)

25

source of electricity but will not begin to compete with the major fossil fuel source plants until costs come down further.

Wrighton's presentation, "Photosynthesis—Real and Artificial," was a closely reasoned, step-by-step discussion of the crucial stages in the chemical and molecular sequence of photosynthesis. His colleagues in the session were chosen for their expertise in one or another of these fundamental specialized areas of photosynthesis research. By the end of the session, they had not only provided a lucid explanation of the process, but had also described firsthand some of the intriguing experimental data produced. Douglas Rees of the California Institute of Technology (on the molecular details of biological photosynthesis), George McLendon of the University of Rochester (on electron transfer), Thomas Mallouk of the University of Texas (on the arrangement of materials to facilitate multielectron transfer chemistry), and Nathan Lewis of the California Institute of Technology (on synthetic systems using liquid junctions) all supplemented Wrighton's overview with reports about findings in their own area of photosynthesis research.

The science of chemistry is predicated on the atomic theory of matter. No matter how intricate the structure of an atom or molecule, its constituent parts will be conserved after the exchanges of a chemical reaction. In fact it was the development of the balance scale in the 18th century that led to the birth of modern chemistry. Once it was realized that the laws of thermodynamics and the principle of the conservation of energy provide an elegant set of constraints, chemistry became the ultimate puzzle-solving science. One could feel fairly confident—once most of the basic elements and compounds and their simple proportional relationships had been discovered—that the answer could be found in the laboratory, if only the pieces could be assembled into the proper, coherent picture. For chemists, this usually means recreating an interaction under conditions that are precisely repeatable.

The enabling paradigm was developed by British chemist John Dalton, who proposed the atomic theory of matter around the turn of the 19th century. Notwithstanding subsequent refinements due to quantum physics and to scientists' increasing ability to probe and examine these reactions directly, Dalton's basic description of the behavior and transfer of protons and electrons among and between elements and compounds—the opening salvo fired at every high school chemistry student—still sets the stage for the most advanced chemical research. Photosynthesis provides a vivid example of the type of drama that is played out effortlessly in nature but reenacted elaborately in chemical laboratories with painstaking concern for the intricate details. It is hardly an oversimplification to say that if scientists could figure out exactly how the electrons liberated by light striking plants are so efficiently put to work in subsequent chemical transformations, analogous systems might well be developed that could help substantially to meet the world's energy needs. Thus the stakes for society are high, and the contrast is dramatic: a chemist works on precisely controlled and intricately choreographed reactions in a laboratory—usually on a very small scale—and yet the implications and applications of his or her work may lead to dramatic transformations, on a vast scale, of the material world.

In the case of research on artificial photosynthesis, such work could lead to the economical production of an alternative to the dwindling supply of fossil fuels. And a further benefit might be a reduction in the sulfurous products emitted by the combustion of carbon-based fuels. Wrighton explained that these fuels are themselves "the result of photosynthetic processes integrated over the ages." It must be kept in mind that long before plants developed the ability to produce molecular oxygen as a byproduct of photosynthesis, they were always about their real business of converting carbon dioxide from the atmosphere into carbohydrates for their own sustenance. In fact some of these anaerobic plants still exist today in certain specialized ecological niches. The system for photosynthesis evolved to its present state during the earth's natural history, and it exploits materials that are naturally abundant and inexpensive, Wrighton pointed out. As designed by nature, it is the ultimate recycling process—since it uses the planet's two most abundant resources, CO_2 and H_2O, providing fuel and breaking down a pollutant.

Wrighton and others in the hunt to develop a synthetic version of the photosynthetic process that would generate commercially viable energy accept these two fundamental provisos: it must be cheap, and its raw material inputs must be abundant. Present PV technology has developed solar efficiencies (that is, a certain percentage of the energy received from the sun deliverable as electricity) of 28.5 percent for point-contact crystalline silicon solar cells and 35 percent for a gallium arsenide–gallium antimonide stacked junction cell (Brodsky, 1990), but the manufacturing costs of these products do not allow them to compete where conventional alternative sources are available. If the two natural photosynthetic inputs, CO_2 and H_2O, could be harnessed, said Wrighton, "fuel mixtures that would be useful within the existing technological framework—where combustion processes dominate the use of our existing fuels"—are foreseeable. One ancillary benefit of such a process could be to "bring global CO_2 concentrations to a steady-state value," manifestly a desirable goal, said Wrighton. But as he pointed out, there are other candidates for an input source that are ubiquitous, including SiO_2 (silicon dioxide in rocks), N_2 and O_2 (molecular nitrogen and oxygen from the air), and NaCl (common table salt).

If one of Earth's abundant natural resources could be energized by sunlight to produce (probably by the breakdown and release of one of its elements) a source that could be used for fuel, the entire fossil fuel cycle and the problems associated with it might be obviated. If that resource were water, for example, and the resultant fuel source were hydrogen, burning liquid hydrogen in the air would produce only water as a combustion product. Liquid hydrogen is already in use as a fuel source and has always been the primary fuel powering space vehicles, since it produces more heat per gram of weight than any other known fuel. If a photosynthetic system delivering usable hydrogen could be developed, the process would regenerate the original water source, and an entirely new recycling of natural resources could be established. This time, however, cultural rather than natural evolution would call the shots. With such a major new fuel production process, science would hopefully be able to provide a methodology to better anticipate and control the global impact of any byproducts or emissions.

The search for a new way to harness the sun's energy involves first the study of how photosynthesis works in nature, and then the attempt to devise a new system that to some extent will probably mimic or model the successful example. Wrighton and his colleagues provided a lucid description of both efforts.

PHOTOSYNTHESIS IN NATURE

Photons and the Physics of Light Energy

Photosynthesis is made possible by the flood of free energy raining down on the planet from the sun. Clearly this energy is put to good use by plants and certain bacteria that have developed analogous photosynthesizing abilities. Many studies of photosynthesis are conducted on these organisms, which are hardy and genetically manipulable under laboratory conditions.

But of what does this energy shower of light consist? How can certain structures convert it to chemical energy that is useful to them? The background to answering this question involves two of the giants of 20th-century physics—Planck and Einstein—whose work at the beginning of this century provided important fundamental insights into the energy of light. German physicist Max Planck in 1900 derived what is called Planck's constant (universally referred to by the symbol h) to demonstrate his theory of the quantization of energy. He realized that the light energy given off by burning bodies varies with the frequency of their vibrating atoms, and produced the formula $E = nvh$, $n = 1, 2, 3, \ldots$.

Thus the energy (E) of a vibrating atom will vary according to its frequency of vibration (v), but can assume only specific quantity values, namely whole integers times h (approximately 6.63×10^{-34} Joule-second). These values, the various products of hv times whole integers, are known as quantum numbers. When one says the energies of atoms are quantized, one means that they can assume values from this set of numbers only. Thus a quantum of energy—whether it be light or electromagnetic energy outside the optical spectrum—provides scientists a measure of the smallest piece of energy that seems to be involved in the transfer events they are trying to understand. "These concepts have been extended to molecules, where absorption of quanta of energy—photons—occurs at particular energy levels and gives rise to molecules that then possess very different properties," and most importantly, Wrighton continued, "they can be useful in effecting oxidation and reduction reactions."

These "particles" of electromagnetic energy were observed to be proportional to the frequency of light in which they were traveling. Thus when a photon of a particular energy strikes a metal, for instance, that metal's outer electron(s) will be ejected by the photoelectric effect only when the incoming photon has sufficient energy to knock it loose. Light and the energy value of the photons it transmits vary according to its wavelength frequency; materials vary according to how easy it is to displace a valence electron. When this does occur, the photon is said to be absorbed by the substance, and actually ceases to exist as a particle. Aerobic plants absorb photons of light from the sun within a certain frequency range, and this drives the movement of electrons that yields the synthesis of carbohydrates and oxygen. This is the theoretical physics underlying photosynthesis. But it is the physical chemistry that interests Wrighton and his colleagues, who hope to develop analogous systems that would produce usable energy.

Harvesting Photons and Putting Them to Use

Two fundamental constraints govern the system: the plant or photosynthesizing organism must possess a mechanism to register or receive the incoming photon; and since the energy content of a single photon is small, a way must also be found to collect and aggregate such energy. Plants have evolved mechanisms to overcome both of these problems. In plants, chlorophyll provides what chemists classify as a sensitizer, a species that absorbs light and effects subsequent chemical reactions. "The assembly of chlorophyll collects and aggregates light energy," Wrighton explained.

As Wrighton pointed out, "First, it is noteworthy that exposure of CO_2 and H_2O to sunlight alone does not lead to a reaction, since the photosynthetic apparatus involves light absorption by molecules other than CO_2 and H_2O, namely chlorophyll. Chlorophyll can therefore be regarded as a sensitizer, a light absorber which somehow channels the light to reaction pathways leading to the production of energy-rich products." It is the absorption by chlorophyll of light frequencies in the lower wavelengths that produces the green color of plants, he explained. But the crucial role played by chlorophyll and any sensitizer is to expand the bandwidth of the energy a system can accept. Wrighton added that since CO_2 and H_2O do not absorb in the visible frequency range, some sort of sensitization will be needed to exploit the full range of the sun's energy in the optical, or visible, range of the electromagnetic spectrum where photon energies tend to be sufficient to dislodge an electron, between 200 and 800 nanometers. This proviso is not limited to the carbon dioxide and water nature breaks down, however, but also applies, said Wrighton, to "all abundant, inexpensive fuel precursors" currently under consideration as candidates for a synthetic system. "Thus, efficient photoconversion systems will involve use of efficient visible light absorbers," he concluded.

"The second critical feature of the photosynthetic apparatus," Wrighton emphasized, is that "in order to achieve high solar conversion efficiency, the formation of a single fuel molecule will involve the energy from more than one photon." The ratio of photons to electrons released will probably be one to one in any system—as it is in nature—but there must be a way to harness the energy from several of these freed-up electrons to produce a chemical transformation. "If a one-photon mechanism were operative in nature, the process would be doomed to low efficiency," he explained, because a single photon that would break down H_2O would have to be in the blue wavelength range, and "sunlight does not contain much energy of blue light, or of shorter wavelengths." Nature's way of aggregating and using more than one photon of the energy that *is* abundant, throughout the entire optical wavelength spectrum, is photosynthesis.

The photosynthesis process in nature involves two separate photosystems (called PS I and PS II), each of which contains the chlorophyll whose electrons provide the vehicle for the energy that is created. The sequence of molecular events occurs within a structure that biochemists classify as the Z-scheme (Figure 2.1). This molecular arrangement accomplishes an oxidation–reduction, or redox, reaction that involves the actual (or in some cases only the apparent) transfer of electrons between species. When these two phenomena occur together, the overall activity is described as a redox reaction, whereby in one half of the reaction a species loses electrons—is oxidized—and in the other half of the reaction a different species gains electrons—is reduced (Ebbing, 1990). Nature uses photons to free electrons from chlorophyll and—through a series of steps—to oxidize H_2O, and in the process O_2 is created as a product of the reduction of CO_2.

The Z-scheme provides an architecture of molecules, located in what biochemists call the reaction center of the plant, that facilitates the redox reaction. Crucial to this arrangement is a mechanism that will serve not only to separate an electron from its atomic species, but will also move it, once it has been separated, in a timed and coordinated way along a known path. Summarizing the three essential elements of the Z-scheme, Wrighton said that the two natural photosystems found in all aerobic plants work in series to (1) absorb four photons of light to energize chlorophyll, (2) release electrons by charge separation and move them by unidirectional charge transport

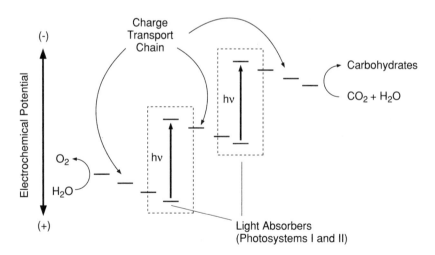

FIGURE 2.1   Z-scheme representation of the photosynthetic apparatus showing, on an electrochemical potential scale, components for light absorption, charge transport, and redox processes for oxidation of H_2O to O_2 and reduction of CO_2 and H_2O to carbohydrates. (Courtesy of M. Wrighton.)

away from the holes they left behind, and then (3) deliver them to sites distant enough from each other to prevent their mutual interference, where their respective oxidizing (of H_2O) and reducing (of CO_2) tasks can be fulfilled.

The fairly astounding concept at the heart of this series of events is that, in order for photosynthesis to occur, evolution had to develop a structure that would facilitate a chemical series of steps that are, even today, reproducible in the laboratory only with great difficulty. Wrote McLendon (1988), "Just a few years ago, the idea of 'long-distance' chemical reactions, in which the reactants are held fixed at distances (10–30 Å) that preclude collisions, was viewed with suspicion or disbelief by many chemists. Fortunately, nature is not so skeptical, and biological energy is channelled in the photosynthetic and respiratory transport chains by just such long-distance reactions."

The Reaction Center

Wrighton reported that "remarkable progress in establishing the molecular structure" of the reaction center has been made in recent years, citing Douglas Rees of the California Institute of Technology as a contributor in this area. There remain some important unanswered questions, however, said Wrighton: Why is the electron transfer from the photoexcited chlorophyll so fast? Why does the electron transfer take only one of two quite similar paths? Rees has focused on "one of the simplest systems for studying biological photosynthetic electron transfer," a system with "only a single photosystem," a type of bacteria that—while it does not produce oxygen or reduce carbon dioxide—nonetheless does photosynthesize, and employs in certain of its cells an electron transfer chain that is comparatively clear to study, and that will most likely yield insights about artificial systems that might be designed and built in the laboratory.

The reaction center molecules are mostly proteins, specialized polypeptides, and strings of amino acid residues. The overall process begins, explained Rees, when "light is absorbed by a specialized pair of the bacterium's chlorophyll molecules." These molecules are called P680 (pigments that absorb light most strongly at the frequency of 680 nanometers) and react to the photon by giving up an electron. Such a structure is then said to be in an excited state and represents the positive component of the charge separation, referred to as a *hole*. As the electron moves through the transport chain, its negative charge and the positively charged hole are separated by ever greater distance, and potential energy is created. In essence, the reaction center acts as a tiny capacitor by separating and storing these positive and negative charges on opposite sides of its architecture, figuratively like the poles of a battery. Four more or less distinct steps constitute the process, after the chlorophyll pigment has absorbed the photon and donated its electron. The moving electron is accepted by the pigment pheophytin very quickly, "in roughly 4 picoseconds," explained Rees, which passes it to a primary quinone, Q_A, and then on to a secondary acceptor quinone, Q_B. Finally, a secondary electron donor gives up an electron to replace the one lost by the original donor, which is thereby reduced (that is, gains an electron). The light energy captured by the reaction center is ultimately utilized to drive the metabolic processes necessary for life.

Much of the detail has been observed directly, said Rees, who pointed out that "crystallographic studies provide a nice framework for rationalizing and understanding the kinetic sequence of events. But also, they raise a number of questions." The most significant of these involves the rates of these various steps. The atomic electrical attraction of positively and negatively charged actors in the process always threatens to draw the liberated electron back into its hole, a process called back electron transfer. If a step proceeds forward at too slow a rate, back transfer will occur and will short-circuit the entire process. In addition to increasing their speed, emphasized McLendon, experimenters also have to steer these freed-up electrons. "It doesn't do any good to move an electron around in a picosecond if it goes to the wrong place, because you will just short-circuit your electron transport chain. Then every cellular component gets to a common free energy, and you have a death by entropy."

Genetic engineering has also been employed to create mutant reaction centers, where certain proteins and cofactors have been deleted or altered. Rees reported that, "rather surprisingly, in many of these mutant forms the reaction center still works," though with a reduced quantum efficiency. In sum, "this marriage of molecular biology and chemical physics has provided a good structural understanding of the reaction center," said Rees, who was also referring to major strides made in spectroscopy, theory, and x-ray crystallography. The dynamics and the energetics of the process still remain imperfectly understood, but, he predicted, "the prognosis looks quite good for unravelling these details in the next 5 years or so."

LABORATORY PHOTOSYNTHESIS: IMPORTANT FIRST STEPS

Wrighton and other chemists around the world are trying to use what has been learned about photosynthesis in nature to create a synthetic system that might produce a renewable source of energy. The energetics Rees referred to have become an important area of inquiry called excited-state electron transfer, advances in which will aid chemists and molecular biologists who are already building actual molecular structures to achieve conversion of light to energy. Thus far, the most promising synthetic systems have exploited the chemistry and physics of liquid-semiconductor junctions.

Excited-state Electron Transfer in Synthetic Systems

Quantum physics explains how the light energy of a photon is absorbed by the right sort of receptor, kicking an electron loose and commencing the process of photosynthesis by creating a source of potential energy between the separated electron and the hole it formerly occupied. Gaining and losing electrons "is the name of the game in redox reactions," said Wrighton, who added, "It has long been known that the photoexcited molecules are both more potent oxidants and more potent reductants than the ground electronic state" (Figure 2.2). When a photon is absorbed to create an electron and a hole, something thermodynamically unstable is produced, and there's always the tendency for the electron and the hole to recombine. Back electron transfer is, metaphorically, a short circuit that bleeds the potential energy of the charge separation before it can aggregate at a distant site within the system and be put to use.

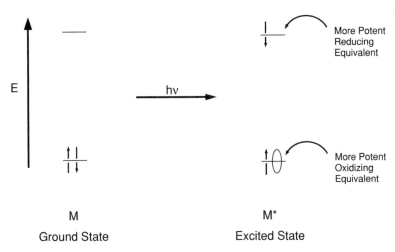

FIGURE 2.2 Orbital diagram rationalizing potent redox power of one-electron excited molecules. (Courtesy of M. Wrighton.)

Wrighton and others have been able to find and to build multi-component molecules with "a donor and an acceptor system covalently linked to a light absorber, an assembly," he pointed out, "that does indeed resemble the heart of the Z-scheme" found in plants. But their system does not produce energy as efficiently as they would like, because of the timing of the reactions: "In solution the energy-rich molecules lose the stored energy by either intermolecular or intramolecular back electron transfer. In nature, the movement of the carriers away from one another is crucial to high efficiency. The unidirectional movement of electrons in the Z-scheme is a consequence of the components of the charge transport chain, and how they are arranged, both geometrically and energetically," Wrighton explained. Work in his group, he continued, did lead to construction of a complex molecule with all of the functional components, but the photoexcitation test showed that a 10-nanosecond time was required for the donor to deliver the molecule to the acceptor. "This is very sluggish compared to the 4 picoseconds demonstrated [in the natural systems]," Wrighton summarized, "and so one of the challenges is to prepare a molecule that will have more zip than our 10-nanosecond time." Thus chemists explore the quantum world, said Wrighton, narrowing in on several factors that might elucidate the transfer rates of the electrons: "the energetics for electrons, the distance dependence of electron transfer, and the structures of the donor and acceptor and their relationship" in space. George McLendon of the University of Rochester, said Wrighton, "has made important progress in understanding such factors."

McLendon is a chemist specializing in the quantum processes of moving electrons from one molecule to another. Not focusing exclusively on photosynthesis, he usually works with proteins and biological systems, but his laboratory has demonstrated phenomena crucial to all electron transfer systems. The basic physics involves the concept of conservation of energy, which, explained McLendon, shows that an electron's rate of transfer varies with the energy force driving it. Essential to the first step in photosynthesis, this relationship between rate and energy was analyzed theoretically some years ago by Rudy Marcus (1956) at Caltech, who predicted an anomaly that was first confirmed by John Miller at Argonne National Laboratory, and verified subsequently by McLendon and others. Up to a certain level of energy, the rate of electron transfer increases with the force driving it, but the initially proportional relationship changes. After the peak level is reached, additional driving force actually slows the electron down. "A funny thing," said McLendon, "is that you can have too much of a good thing."

Quantum physics provides a suggestive, if incomplete, explanation of why the electron does not transfer ever more quickly as more energy is applied. McLendon explained that when an electron is lost from one molecule (a donor) and gained by another (an acceptor), the lengths of the bonds in each molecule change, like a stretching spring. The energy for this stretching, called the "reorganization energy," said McLendon, "is supplied by the excess [free] energy of reaction. The fastest rate occurs when the reaction free energy exactly equals the reorganization energy. If less energy is available, the bonds can't be stretched enough for reaction to occur." Conversely, continued McLendon, "if too much energy is available, the system must wait for this extra energy to 'dribble' away, since at the instant of transfer, the energy of the starting materials and products must be equal, by the law of conservation of energy."

McLendon provided an example: "Consider a system containing only ferrous ions (dipositive charged ion molecules) and ferric (tripositive) ions. When an electron moves, one ferrous becomes ferric, and vice versa. Obviously the system, as a whole, remains unchanged. However, in each case the molecules' bonds *have* changed *length*—ferric bonds are shorter. If this length change occurred *after* the electron moved, energy would be released. With each subsequent electron transfer, the solution would become warmer, and there would be no [so-called] energy crisis. Obviously," continued McLendon, "nature *doesn't* work this way. Instead, heat is first taken from the system to stretch the bonds. Then, the electron jumps. Finally, the bonds relax, and heat is given back up. Overall, there is no change in heat, and energy is conserved. However, the slow step is getting the initial energy to stretch the bonds. If this can be provided by the energy of reaction, the reaction rate increases." The dependence of the separation rate on energy is approximately exponential, falling away from the peak value at lower, or higher, energies. Marcus' theoretical prediction of this phenomenon his colleagues believe was prescient, and he has continued to influence developments in the study of reorganizational energy (Marcus and Sutin, 1985).

Lessons for and from the Construction of Photodiodes

Thomas Mallouk, a chemistry professor at the University of Texas, Austin, wants to make certain that such theoretical insights are put to use. He sees an object lesson in the history of research on superconductors in the 1950s, when increasing knowledge led only very slowly to any "perturbation in the technology." So at his laboratory chemists are actually trying to build a photodiode from which to obtain usable electric current. Using insights from "the two systems that really work—photosynthesis and photovoltaics that use semiconductors—the first question in creating a multistep electron transfer mechanism," according to Mallouk, "is, What is the best way to organize the molecules?"

Because the cost of materials and the complexity of assembly are crucial determinants of the viability of a commercial system, Mallouk said, "we look for ways where we can get these things to self-assemble, hopefully into a matrix that will teach them how to line up usefully, even if we just chuck them all in there randomly." Several strategies are combined, including using electric field gradients to propel and direct freed electrons, creating a gate whose size will selectively admit only the desired species of molecule, and employing molecules whose electrochemical potentials are easily controlled. (Figures 2.3 and 2.4).

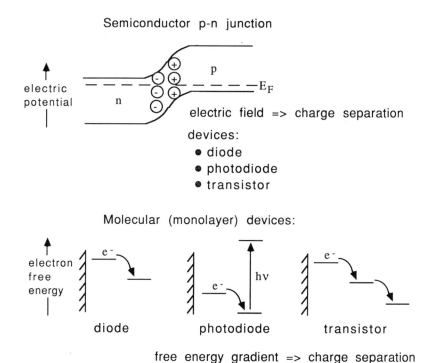

FIGURE 2.3 Analogy between semiconductor- and molecule-based electronic devices. (Courtesy of T.E. Mallouk.)

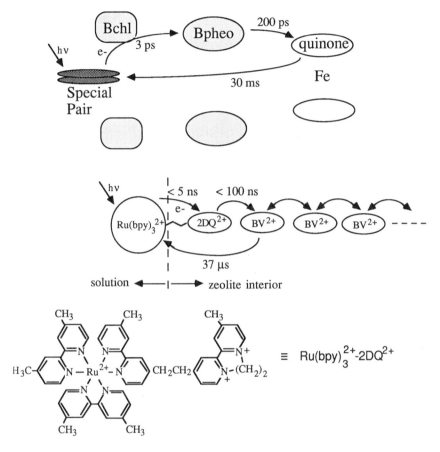

FIGURE 2.4 Analogy between natural and zeolite-based photosynthetic reaction centers. (Courtesy of T.E. Mallouk.)

Zeolites are a form of aluminosilicate. "There are many different polymorphs of this general formula," said Mallouk, "and they have the endearing property that the holes in them are about the size of a molecule, about 6 or 7 angstroms across. You have an array of linear tunnels that run through the structure, with the tunnels packed in next to one another." By exploiting the diameter of the tunnel to exclude molecules of a certain size, adding some useful catalysts, and charging the system with a calculated amount of energy, Mallouk's team was able to trap the target molecule inside the tunnel, where it could not escape while the scientists beamed their light and photolyzed it. The first transfer step occurs when the electron escapes from a molecule in solution to interact with molecules that have been chemically attracted to the surface of the assembly; the second step involves the product of that step reacting with molecules inside the tunnel altogether. "So we have a self-assembling, three-molecule chain, where the viologen is inside, porphyrin is on the outside, and EDTA is ion-paired with the porphyrin out there," Mallouk explained. The assembly worked, in that it propelled the electron into the tunnel in 30 picoseconds, only 10 times slower than with natural photosynthesis. The efficiency of the electron-hole separation was very poor, however, said Mallouk, illustrating Wrighton's emphasis on the importance of back electron transfer pathways.

"We had to think more intelligently for our Mach II version," said Mallouk, who then added two more components to the system in order to better mimic nature. "Remember, the natural system is a sort of staircase arrangement of redox potentials," he said, and so they tried to create a multistep mechanism within their molecular assembly. Rather than speed up the process, they chemically blocked the back transfer pathway by manipulating the cationic charges and using the 7.1-angstrom-wide zeolite tunnel as a kind of trap. Their molecule was electrically propelled through, small-end first, but then became trapped like a pig in a fence as its other larger end reached the tunnel. Mallouk's team actually "slowed down the forward rate," which, he conceded, is "kind of a macho thing in electron transfer chemistry, where the basic rule is to get your forward rates as fast as possible." But, he said, citing McLendon's thesis, "it really doesn't matter as long as your forward rate is fast enough compared to the competing back transfer pathway."

"There are many steps in a row, but the point is to get the electron safely away from where it came. With so much empty space in the zeolite, we can load up the rest of the channel with a molecule that is easier to reduce. Now we have two steps forward, which chemically puts the electron much further away from its hole, so the rate back down [the hill] decreases by a factor of 100 or so," Mallouk explained. How did it work? He continued, "The effect of putting in these more intelligent features is to increase the quantum efficiency by about four orders of magnitude. When you put this whole thing together, you really only have to stir it up, and you come out with a quantum yield of about 30 percent for photochemical separation of electrons and holes, and subsequent hydrogen evolution." Thus, Mallouk concluded, they have created another system—this one man made—that works, and that delivers electrons productively. However, even though his system has the advantage of self-assembly, Mallouk noted that it still employs components that are expensive to create.

Although such systems may eventually evolve into competitive models, Wrighton and his team are currently at work on a prototype that will be compared to, he said, "the system to beat, the ones that work very well, semiconductor-based photoelectrochemical cells."

PHOTOVOLTAIC CELLS

As Wrighton said, not only does photosynthesis *work*—in nature and in the laboratory—but it can become commercially viable. Solar technology has been a growing phenomenon for two decades. The road to this growth, however, is really a two-lane highway, a fact that is often overlooked. Any alternative system that can produce electricity will be measured by how it compares to the current massive combustion of fossil fuels. On the one hand, scientists, ecologists, and even policymakers taking the long view recognize the finite and dwindling supply of these carbon-based sources and the harmful impacts of their sulfurous emissions. On the other hand, such considerations tend to be overshadowed by more immediate concerns, such as the cost of energy produced by alternative systems and how quickly that energy can be delivered. Market considerations, pragmatic politics, and the interests of the massive infrastructure now producing and selling electricity all deflect attention and, inevitably, resources away from the support of scientific efforts to develop a viable alternative source of energy.

Solar energy itself is multifaceted. Without the sun to evaporate the water that eventually returns to the earth's rivers and oceans, hydroelectric power would not have achieved the importance it has, providing approximately 17 percent of the energy used in industrialized countries and 31 percent in the developing countries. California's Altamont Pass is the site of the world's largest wind farm, indicating the feasibility and possibility of wind turbines powered by the earth's atmosphere, which in turn, of course, is powered by the sun. In addition, solar cells have been designed to collect heat and transfer it directly into heating and cooling uses at the site, as well as to heat a fluid that can be used to generate power in a system known as thermal electric generation. But the "system to beat" referred to by Wrighton, photovoltaic cells, creates energy with no noise and no pollution, and does so in a self-contained unit whose only drawback at the moment is cost of fabrication. If the design of photoelectrochemical cells could be improved to make the cost per kilowatt-hour of their generated electricity competitive, or to yield a fuel product from ubiquitous raw materials, the world's energy future could be reconsidered altogether.

Construction of Typical Solar Cells

A typical solar cell is made from two different types of semiconductor, each having a characteristic—and opposite—electrical tendency when perturbed by light. So-called *p*-type semiconductors, said Wrighton, "exhibit conductivity by virtue of having positively charged carriers, *n*-type semiconductors by virtue of having negatively charged carriers." The typical solar cell looks like a sandwich, engineered to exploit these preferences or atomic tendencies inherent in the silicon or other material of which the usually metallic semiconductor is fabricated. The base of the cell is a substrate of glass or plastic that insulates the cell. Above this cushion is the first of the two (outside, top and bottom) metal contacts that will eventually deliver the current in one direction each. Next comes the first of two silicon layers—say the *p*-type, designed to create and attract holes—and then an interface where an electric field is built into the cell as a type of particle accelerator. Above this is the second semiconductor, the *n*-type, through most of the upper layers of which photons from the sun will penetrate from above. This layer, including an antireflective coating at its upper edge, is designed so that photons moving through the material will be absorbed as close as possible to the magnetic interface with the other semiconductor. Thus, when an electron is released and creates a hole, each of these particles will be maximally positioned to begin their respective journeys through the respective semiconductors for which they have an inherent electrical affinity. Propelled by the electric field and their own charge, they migrate through the semiconductor layers to the metal contacts at top and bottom, from which a current can be tapped.

Use of Liquid Semiconductor Junctions

Nathan Lewis, now at Caltech, did his early work with Wrighton at MIT. Each of them is now working on a version of the photovoltaic (PV) cell described above, but with an important difference. Instead of a semiconductor-to-metal or solid-to-solid interface between the two particle-generating media, they are exploiting a liquid-semiconductor junction. "There are many important practical advantages," said Wrighton, who conceded that one way of looking at the system is "as a solid-state PV device [arranged electrically] in series with an electrolysis cell. . . . The junction is easily prepared by simply immersing the semiconductor into the liquid electrolyte." Reminding the symposium's scientists that the payoff involves redox chemistry, Wrighton said that "further, the semiconductor surface may be unique for certain redox processes that would be chemically inhibited on conventional metallic electrodes. . . . Most significant is the prospect that the liquid junction device could yield higher efficiencies than devices based on p- or n-semiconductor junctions." Added Lewis: "Really it is a cost-performance analysis. The physics of recombination can rigorously be demonstrated to be superior to that at the best semiconductor–semiconductor interfaces. We can control the interfacial chemistry."

The reason has to do with quantum physics at the surfaces. If the semiconductor is in contact with a metal, said Lewis, "you find that the resultant voltages are less than thermodynamics analysis predicts," because of interfacial reactions. A similar problem plagues the interface between two semiconductors of different electrical characters. "If you try to match their surfaces with atomic precision, you will pay a price" to do so, said Lewis, and thus drive up the economic cost of the system. "When you miss at certain spots, those spots become recombination sites," and some of the free charge meant to be collected as electricity is drained into these surface reactions. Using a liquid at the interface obviates both of these problems. First, one can add into the liquid something else that will have a high affinity for the defective sites that could have led to problematic recombination, and can thereby passivate these sites. Second, one can reduce the danger of back electron transfer by choosing a solvent that draws electrons more strongly and increases their forward rate.

There is a trade-off, however, because the solvent used in the basic solution determines the wavelength of light necessary to kick loose an electron. Water, it turns out, oxidizes silicon, essentially "poisons it," said Lewis. "You can suppress that reaction by going to a nonaqueous solvent," he continued, "but you then surrender a key property, the ability to extract simple fuels of H_2 and O_2." For water, Lewis and his co-workers substituted an iron-containing molecule (ferrocene), an ideal donor to accept the holes generated as light kicks loose the silicon electrons. By carefully mixing their solution, Lewis and his team were able to produce a solvent that maximized the forward rate of electron transfer. The advantage of their cell, Lewis pointed out, is that it has an efficiency in the sun of over 14 percent and has been shown to be stable for months. However, it is not capable of simultaneously producing chemical fuel. Lewis sees the challenge as three-fold: achieving stability, efficiency, and simultaneous fuel formation, two-thirds of which has been demonstrated in the silicon cell.

Wrighton described the strontium-titanate cell as "extraordinarily efficient for the wavelengths of light that excite strontium titanate," the downside from the practical perspective being that those wavelengths are shorter and "fall in the ultraviolet range, which contains only 5 percent of the available solar energy." Thus, like other experimental systems and cells, Wrighton's assembly works impressively under laboratory conditions, but it uses an energy source (or in other cases, molecular components) that nature does not have an abundance of. "Short of changing the sun itself," said Wrighton, the problem remains to find a stable solution that will provide a complete cycle of oxidation and reduction, and a donor to complement it that will react to visible light. A real frontier in research on PV cells is finding or constructing materials and solvents that will balance all of these considerations: bandgap, passivation to counter deficits at the surface junctions, ionic acceleration of particles to avoid back transfer, and a material that will provide both sides of the redox reaction so as to produce a usable fuel.

Wrighton summarized the challenge: "Although the semiconductor-based photoelectrochemical cells represent the best chemical systems for conversion of light to electricity or chemical fuel, there are many shortcomings. Long-term durability for any chemical system remains a serious question. Nonetheless, the $SrTiO_3$-semiconductor-based cell for photoelectrolysis of H_2O remains as a good demonstration that sustained conversion of light to useful energy is possible. In trying to duplicate the *function* of the natural photosynthetic apparatus," he continued, "semiconductor-based approaches are far ahead of molecular-based approaches. . . . But before we see a mature solar conversion technology based on the excitation of electrons in materials, advances in chemistry and materials science must overcome present obstacles to efficient, large-area, inexpensive systems."

BIBLIOGRAPHY

Brodsky, Marc H. 1990. Progress in gallium arsenide semiconductors. Scientific American 262(February):68-75.

Ebbing, Darrell D. 1990. General Chemistry. Third edition. Houghton Mifflin, Boston.

Grätzel, Michael. 1990. Artificial photosynthesis. Pp. 83-96 in Frontiers of Science. Andrew Scott (ed.). Basil Blackwell, Oxford.

McLendon, George. 1988. Long-distance electron transfer in proteins in model systems. Accounts of Chemical Research 21(April):160-167.

Marcus, R.A. 1956. On the theory of oxidation–reduction reactions involving electron transfer. Journal of Chemical Physics 24:966-978.

Marcus, R.A., and Norman Sutin. 1985. Electron transfers in chemistry and biology. Biochimica et Biophysica Acta 811:265-322.

RECOMMENDED READING

Gust, D., and T.A. Moore. 1989. Mimicking photosynthesis. Science 244:35-41.

Krueger, J.S., C. Lai, Z. Li, J.E. Mayer, and T.E. Mallouk. 1990. Artificial photosynthesis in zeolite-based molecular assemblies. Pp. 365-378 in Inclusion Phenomena and Molecular Recognition. J.L. Atwood (ed.). Plenum Press, New York.

3

When the Simple Is Complex: New Mathematical Approaches to Learning About the Universe

The science of dynamical systems, which studies systems that evolve over time according to specific rules, is leading to surprising discoveries that are changing our understanding of the world. The essential discovery has been that systems that evolve according to simple, deterministic laws can exhibit behavior so complex that it not only appears to be random, but also in a very real sense cannot be distinguished from random behavior.

Dynamical systems is actually an old field, because since ancient times we have tried to understand changes in the world around us; it remains a formidable task. Understanding systems that change linearly (if you double the displacement, you double the force), is not so difficult, and linear models are adequate for many practical tasks, such as building a bridge to support a given load without the bridge vibrating so hard that it shakes itself apart.

But most dynamical systems—including all the really interesting problems, from guidance of satellites and the turbulence of boiling water to the dynamics of Earth's atmosphere and the electrical activity of the brain—involve nonlinear processes and are far harder to understand.

Mathematician John Hubbard of Cornell University, lead speaker in the dynamical systems session of the Frontiers symposium, sees the classification into linear and nonlinear as "a bit like classifying

---

NOTE: Addison Greenwood collaborated on this chapter with Barbara A. Burke.

people into friends and strangers." Linear systems are friends, "occasionally quirky [but] essentially understandable." Nonlinear systems are strangers, presenting "quite a different problem. They are strange and mysterious, and there is no reliable technique for dealing with them" (Hubbard and West, 1991, p. 1).

In particular, many of these nonlinear systems have been found to exhibit a surprising mix of order and disorder. In the 1970s, mathematician James Yorke from the University of Maryland's Institute for Physical Science and Technology used the word *chaos* to describe the apparently random behavior of a certain class of nonlinear systems (York and Li, 1975). The word can be misleading: it refers not to a complete lack of order but to apparently random behavior with nonetheless decipherable pattern. But the colorful term captured the popular imagination, and today *chaos* is often used to refer to the entire nonlinear realm, where randomness and order commingle.

Now that scientists are sensitive to the existence of chaos, they see it everywhere, from strictly mathematical creations to the turbulence in a running brook and the irregular motions of celestial bodies. Nonrigorous but compelling evidence of chaotic behavior in a wide array of natural systems has been cited: among others, the waiting state of neuronal firing in the brain, epidemiologic patterns reflecting the spread of disease, the pulsations of certain stars, and the seasonal flux of animal populations. Chaotic behavior has even been cited in economics and the sociology of political behavior.

To what extent the recognition of chaos in such different fields will lead to scientific insights or real-world applications is an open question, but many believe it to be a fundamental tool for the emerging science of complexity.

Certainly the discovery of "chaos" in dynamical systems—and the development of mathematical tools for exploring it—is forcing physicists, engineers, chemists, and others studying nonlinear processes to rethink their approach. In the past, they tended to use linear approximations and hope for the best; over the past two or three decades they have come to see that this hope was misplaced. It is now understood that it is simply wrong to think that an approximate model of such a system will tell you more or less what is going on, or more or less what is about to happen: a slight change or perturbation in such a system may land you somewhere else entirely.

Such behavior strikes at the central premise of determinism, that given knowledge of the present state of a system, it is possible to project its past or future. In his presentation Hubbard described how the mathematics underlying dynamical systems relegates this deterministic view of the world to a limited domain, by destroying confidence in the predictive value of Newton's laws of motion. The laws are not wrong, but they have less predictive power than scientists had assumed.

On a smaller scale, dynamical systems has caused some upheavals and soul-searching among mathematicians about how mathematics is done, taught, and communicated to the public at large. (All participants in the session were mathematicians: William Thurston from Princeton University, who organized it; Robert Devaney of Boston University; Hubbard; Steven Krantz of Washington University in St. Louis; and Curt McMullen from Princeton, currently at the University of California, Berkeley.)

Mathematicians laid the foundations for chaos and dynamical systems; their continuing involvement in such a highly visible field has raised questions about the use of computers as an experimental tool for mathematics and about whether there is a conflict between mathematicians' traditional dedication to rigor and proof, and enticing color pictures on the computer screen or in a coffee-table book.

"Traditionally mathematics tends to be a low-budget and low-tech and nonslick field," Thurston commented in introducing the session. In chaos, "you don't necessarily hear the theorems as much as see the pretty pictures," he continued, "and so, it is the subject of a lot of controversy among mathematicians."

THE MATHEMATICS OF DYNAMICAL SYSTEMS

Historical Insights—Implications of Complexity

The science of dynamical systems really began with Newton, who first realized, in Hubbard's words, "that the forces are simpler than the motions." The motions of the planets are complex and individual; the forces behind those motions are simple and universally applicable. This realization, along with Newton's invention of differential equations and calculus, gave scientists a way to make sense of the world around them. But they often had to resort to simplifications, because the real-world problems studied were so complex, and solving the equations was so difficult—particularly in the days before computers!

One problem that interested many scientists and mathematicians—and still does today—is the famous "three-body" problem. The effect of two orbiting bodies on each other can be calculated precisely. But what happens when a third body is considered? How does this perturbation affect the system?

In the late 1880s the king of Sweden offered a prize for solving this problem, which was awarded to French mathematician and physicist Henri Poincaré (1854–1912), a brilliant scientist and graceful writer with an instinct for delving beneath the surface of scientific problems.

In fact, Poincaré did not solve the three-body problem, and some of his ideas were later found to be wrong. But in grappling with it, he conceived many major insights into dynamical systems. Among these was the crucial realization that the three-body problem was in an essential way unsolvable, that it was far more complex and tangled than anyone had imagined. He also recognized the fruitfulness of studying complex dynamics qualitatively, using geometry, not just formulas.

The significance of Poincaré's ideas was not fully appreciated at the time, and it is said that even Poincaré himself came to doubt their implications. But a number of mathematicians followed up the many leads he developed. In France immediately after World War I, Pierre Fatou (1878–1929) and Gaston Julia (1893–1978) discovered and explored what is now known as Julia sets, seen today as characteristic examples of chaotic behavior. In the United States George Birkhoff developed many fundamental techniques; in 1913 he gave the first correct proof of one of Poincaré's major conjectures.

Although dynamical systems was not in the mainstream of mathematical research in the United States until relatively recently, work continued in the Soviet Union. A major step was taken in the 1950s, by the Russian Andrei Kolmogorov, followed by his compatriot Vladimir Arnold and German mathematician Jürgen Moser, who proved what is known as the KAM (Kolmogorov-Arnold-Moser) theorem.

They found that systems of bodies orbiting in space are often astonishingly stable and ordered. In a book on the topic, Moser wrote that Poincaré and Birkhoff had already found that there are "solutions which do not experience collisions and do not escape," even given infinite time (Moser, 1973, p. 4). "But are these solutions exceptional?" he asked. The KAM theorem says that they are not.

In answering that question, the subtle and complicated KAM theorem has a great deal to say about when such systems are stable and when they are not. Roughly, when the periods, or years, of two bodies orbiting a sun can be expressed as a simple ratio of whole numbers, such as 2/5 or 3—or by a number close to such a number—then they will be potentially unstable: resonance can upset the balance of gravitational forces. When the periods cannot be expressed as such a ratio, the system will be stable.

This is far from the whole story, however. The periods of Saturn and Jupiter have the ratio 2/5; those of Uranus and Neptune have a ratio close to 2. (In fact, some researchers in the 19th century, suspecting that rationality was associated with instability, speculated that a slight change in the distance between Saturn and Jupiter would be enough to send Saturn shooting out of our solar system.)

But the KAM theorem, combined with Birkhoff's earlier work, showed that there are little windows of stability within the "unstable zones" associated with rationality; such a window of stability could account for the planets' apparent stability. (The theorem also answers questions about the orbit of Saturn's moon Hyperion, the gaps in Saturn's rings, and the distribution of asteroids; it is used extensively to understand the stability of particles in accelerators.)

In the years following the KAM theorem the emphasis in dynamical systems was on stability, for example in the work in the 1960s by topologist Stephen Smale and his group at the University of California, Berkeley. But more and more, researchers in the field have come to focus on instabilities as the key to understanding dynamical systems, and hence the world around us.

A key role in this shift was played by the computer, which has shown mathematicians how intermingled order and disorder are: systems assumed to be stable may actually be unstable, and apparently chaotic systems may have their roots in simple rules. Perhaps the first to see evidence of this, thanks to a computer, was not a mathematician but a meteorologist, Edward Lorenz, in 1961 (Lorenz, 1963). The story as told by James Gleick is that Lorenz was modeling Earth's atmosphere, using differential equations to estimate the impact of changes in temperature, wind, air pressure, and the like. One day he took what he thought was a harmless shortcut: he repeated a particular sequence but started halfway through, typing in the midpoint output from the previous printout—but only to the three decimal places displayed on his printout, not the six decimals calculated by the program. He then went for a cup of coffee and returned to find a totally new and dramatically altered outcome. The small change in initial conditions—figures to three decimal places, not six—produced an entirely different answer.

Sensitivity to Initial Conditions

Sensitivity to initial conditions is what chaos is all about: a chaotic system is one that is sensitive to initial conditions. In his talk McMullen described a mathematical example, "the simplest dynamical system in the quadratic family that one can study," iteration of the polynomial $x^2 + c$ when $c = 0$.

To *iterate* a polynomial or other function, one starts with a number, or seed, x_0, and performs the operation. The result (called x_1) is used as the new input as the operation is repeated. The new result (called x_2) is used as the input for the next operation, and so on. Each time the output becomes the next input. In mathematical terminology,

$$x_1 = f(x_0), \quad x_2 = f(x_1), \quad x_3 = f(x_2), \ldots$$

In the case of $x^2 + c$ when $c = 0$ (in others words, x^2), the first operation produces x_0^2, the second produces $(x_0^2)^2$, and so on. For $x = 1$, the procedure always results in 1, since 1^2 equals 1. So 1 is called a fixed point, which does not move under iteration. Similarly, -1 jumps to 1 when squared and then stays there.

What about numbers between -1 and 1? Successive iterations of any number in that range will move toward 0, called an "attracting fixed point." It does not matter where one begins; successive iterations of any number in that range will follow the same general path as its nearby neighbors. These "orbits" are defined as stable; it does not matter if the initial input is changed slightly. Similarly, common sense suggests, and a calculator will confirm, that any number outside the range $-1 \leq x \leq 1$ will iterate to infinity.

The situation is very different for numbers clustered around 1 itself. For example, the number 0.9999999 goes to 0 under iteration, whereas the number 1.0000001 goes to infinity. One can get arbitrarily close to 1; two numbers, almost equal, one infinitesimally smaller than 1 and the other infinitesimally larger, have completely different destinies under iteration.

If one now studies the iteration of $x^2 + c$, this time allowing x and c to be complex numbers, one will obtain the Julia set (Figure 3.1), discovered by Fatou and Julia in the early 20th century, and first explored with computers by Hubbard in the 1970s (Box 3.1). (To emphasize that we are now discussing complex numbers, we will write $z^2 + c$, z being standard notation for complex numbers.)

"One way to think of the Julia set is as the boundary between those values of z that iterate to infinity, and those that do not," said McMullen. The Julia set for the simple quadratic polynomial $z^2 + c$ when $c = 0$, is the unit circle, the circle with radius 1: values of z outside the circle iterate to infinity; values inside go to 0.

Thus points arbitrarily close to each other on opposite sides of the unit circle will have, McMullen said, "very different future trajectories." In a sense, the Julia set is a kind of continental divide (Box 3.2).

Clearly, to extrapolate to real-world systems, it behooves scientists to be aware of such possible instabilities. If in modeling a sys-

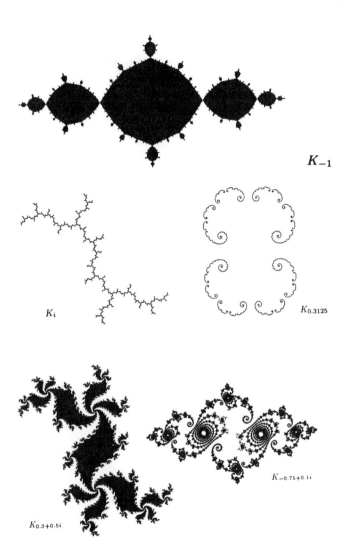

FIGURE 3.1 Typical Julia sets in the z-plane (the plane of complex numbers), showing the points that do not go to infinity on iteration. (Reprinted with permission from Hubbard and West, 1991. Copyright © 1991 by Springer-Verlag New York Inc.)

BOX 3.1   COMPLEX NUMBERS

To define the Julia set of the polynomial $x^2 + c$ (the Julia set being the boundary between those values of x that iterate to infinity under the polynomial, and those that do not), one must allow x to be a *complex* number.

Ordinary, *real* numbers can be thought of as points on a number line; as one moves further to the right on the line, the numbers get bigger. A complex number can be thought of as a point in the plane, specified by two numbers, one to say how far left or right the complex number is, the other to say how far up or down it is.

Thus we can write a complex number z as $z = x + iy$, where x and y are real numbers, and i is an imaginary number, the square root of -1. Traditionally the horizontal axis represents the real numbers, and so the number x tells how far right or left z is.   The vertical axis represents imaginary numbers, so the number y tells how far up or down z is. To plot the complex number $2 + 3i$, for example, one would move two to the right (along the real axis) and up three. (Of course, real numbers are also complex numbers: the real number 2 is the complex number $2 + 0i$.)

Using this notation and the fact that $i^2 = -1$, complex numbers can be added, subtracted, multiplied, and so on. But the easiest way to visualize multiplying complex numbers is to use a different notation. A complex number can be specified by its radius (its distance from the origin) and its polar angle (the angle a line from the point to the origin forms with the horizontal axis.)   The radius and angle are known as polar coordinates. Squaring a complex number squares the radius and doubles the angle.

To return now to the polynomial $z^2 + c$ when $c = 0$, we can see that if we take any point on the unit circle as the beginning value of z, squaring it repeatedly will cause it to jump around the circle (the radius, 1, will remain 1 on squaring, and the angle will keep doubling). Any point outside the unit circle has radius greater than 1, and repeatedly squaring it will cause it to go to infinity. Thus the Julia set for $z^2 + c$ when $c = 0$ is the unit circle.

BOX 3.2 THE FEIGENBAUM-CVITANOVIC CASCADE

Another viewpoint on chaos is given by the Feigenbaum-Cvitanovic *cascade of period doublings,* which, in McMullen's words, "in some sense heralds the onset of chaos." It shows what happens to certain points *inside* Julia sets.

While by definition points outside the Julia set always go to infinity on iteration, points inside may have very complicated trajectories. In the simplest case of $z^2 + c$, when $c = 0$, the points inside all go to 0, but nothing so simple is true for all Julia sets.

Figure 3.2 shows what happens in the long run as you iterate $x^2 + c$ for $-2 \leq c < 0.25$ and x real. To draw it, you iterate each function $f(x) = x^2 + c$, starting at $x = 0$. The first iterate is $0^2 + c = c$, so that on iteration c is used for the next x value, giving $c^2 + c$, which is used for the next x value, and so on.

For each value of c from 0.25 to -2 perform, say, 100 iterations without marking anything, and then compute another 50 or so, this time marking the value you get after each iteration (using one axis to represent values of x and the other for values of c).

The picture that emerges is the cascade of period doublings. First, a single line descends in a curve, then it splits into two curves, then— sooner this time—those two curves each split into two more, and— sooner yet—these four split into eight, and so on. Finally, at about $c = -1.41$ you have a chaotic mess of points (with little windows of order that look just like miniature cascades).

What does the picture represent? The single line at the top is formed of points that are attracting fixed points for the polynomials $x^2 + c$: one such fixed point for each value of c. Where the line splits, the values produced by iteration bounce between two points (called an attractive cycle of period two); where they split again they bounce between four points (an attractive cycle of period four), and so on.

Cascades of period doublings have been observed in nature, said McMullen. Moreover, a remarkable feature is that the pace at which the period doublings occur, first slowly, then picking up speed, as they come closer and closer together, can be represented by a number that is the same whatever system one is studying.

"This is a universal constant of nature," McMullen said. "It is a phenomenon that may have remarkable implications for the physical sciences."

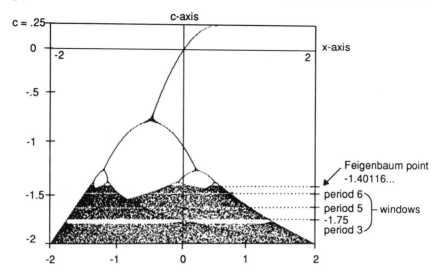

FIGURE 3.2 The cascade of period doublings, heralding the onset of chaos. It is produced by iterating $x^2 + c$, for each value of c. The single line at the top is formed of points that are attracting fixed points. Where the line splits, the values produced by iteration bounce between two points; where they split again they bounce between four points, and so on. (Reprinted with permission from Hubbard and West, 1991. Copyright © 1991 by Springer-Verlag New York Inc.)

tem one is in the equivalent of the stable range, then approximate models work. If one is, so to speak, close to the Julia set, then a small change in initial conditions, or a small error in measurement, can change the output totally (Figure 3.2).

CHAOS, DETERMINISM, AND INFORMATION

Instabilities and Real-world Systems

In his talk, "Chaos, Determinism, and Information," Hubbard argued that this is a very real concern whose ramifications for science and philosophy have not been fully appreciated. He began by asking, "Given the limits of scientific knowledge about a dynamical system in its present state, can we predict its behavior precisely?" The answer, he said, is no, because "instabilities in such systems amplify the inevitable errors."

Discussing ideas he developed along with his French colleague Adrien Douady of the University of Paris-XI (a joint article has been published in the *Encyclopedia Universalis*), Hubbard asked participants whether they thought it conceivable that a butterfly fluttering its wings someplace in Japan might in an essential way influence the weather in Ithaca, New York, a few weeks later. "Most people I approach with this question tell me, 'Obviously it is not true,'" Hubbard explained, "but I am going to try to make a serious case that you can't dismiss the reality of the butterfly effect so lightly, that it has to be taken seriously."

Hubbard began by asking his audience to consider a more easily studied system, a pencil stood on its point. Imagine, he said, a pencil balanced as perfectly as possible—to within Planck's constant, so that only atomic fluctuations disturb its balance. Within a bit more than a tenth of a second the initial disturbance will have doubled, and within about 4 seconds—25 "doubling times"—it will be magnified so greatly that the pencil will fall over.

As it turns out, he continued, the ratio of the mass of a butterfly to the mass of the atmosphere is about the same as the ratio of the mass of a single atom to the mass of a pencil. (It is this calculation, Hubbard added, that for the first time gave him "an idea of what Planck's constant really meant. . . . to determine the position of your pencil to within Planck's constant is like determining the state of the atmosphere to within the flutterings of one butterfly.")

The next question is, how long does it take for the impact of one butterfly on the atmosphere to be doubled?

> In trying to understand what the doubling time should be for equations determining the weather, I took the best mathematical models that I could find, at the Institute for Atmospheric Studies in Paris. In their model, the doubling time is approximately 3 days. Thus, roughly speaking, if you have completely lost control of your system after 25 doubling times, then it is very realistic to think that the fluttering of the wings of a butterfly can have an absolutely determining effect on the weather 75 days later.

"This is not the kind of thing that one can prove to a mathematician's satisfaction," he added. "But," he continued, "it is not something that can just be dismissed out of hand." And if, as he believes, it is correct, clearly "any hope of long-term weather prediction is absurd."

From the weather Hubbard proceeded to a larger question. Classical science, since Galileo, Newton, and Laplace, has been predicated on deterministic laws, which can be used to predict the future or deduce the past. This notion of determinism, however, appears incompatible with the idea of free will. The contradiction is "deeply disturbing," Hubbard said. "Wars were fought over that issue!"

The lesson of a mathematical theorem known as the shadowing lemma is that, at least for a certain class of dynamical systems and possibly for all chaotic systems, it is impossible to determine who is right, a believer in free will or a believer in deterministic, materialistic laws.

In such systems, known as hyperbolic dynamical systems, uncertainties double at a steady rate until the cumulative unknowns destroy all specific information about the system. "Any time you have a system that doubles uncertainties, the shadowing lemma of mathematics applies," Hubbard said. "It says that you cannot tell whether a given system is deterministic or randomly perturbed at a particular scale, if you can only observe it at that scale."

The simplest example of such a system, he said, is angle doubling. Start with a circle with its center at 0, and take a point on that circle, measuring the angle that the radius to that point forms with the x-axis. Now double the angle, double it again and again and again. You know the starting point exactly and you know the rules, and so of course you can predict where you will be after, say, 50 doublings.

But now imagine that each time you double the angle you jiggle it "by some unmeasurably small amount," such as 10 to 15 (chosen because that is the limit of accuracy for most computers: they cut off numbers after 15 decimal digits). Now "the system has become completely random in the sense that there is nothing you could measure at time zero that still gives you any information whatsoever after 50 moves."

What is unsettling—and this is the message of the shadowing lemma—is that if you were presented with an arbitrarily long list of figures (each to 15 decimals) produced by an angle-doubling system, you could not tell whether you were looking at a randomly perturbed system or an unperturbed (deterministic) one. The data would be consistent with both explanations, because the shadowing lemma tells us that (again, assuming you can measure the angles only to within 10 to 15 degrees) there always is a starting point that, under an unperturbed system, will produce the same data as that produced by a randomly perturbed system. Of course it will not be the same starting point, but since you do not know the starting point, you will not be able to choose between starting point a and a deterministic system, and some unknowable starting point b and a random one.

"That there is no way of telling one from the other is very puzzling to me, and moreover there is a wide class of dynamical systems to which this shadowing lemma applies," Hubbard said. Scientists modeling such physical systems may choose between the deterministic differential equation approach, or a probabilistic model; "the facts are consistent with either approach."

"It is something like the Heisenberg-Schrödinger pictures of the universe" in quantum mechanics, he added. "In the Heisenberg picture you think that the situation of the universe is given for all time. . . . You are just discovering more and more about it as time goes on. The Schrödinger picture is that the world is really moving and changing, and you are part of the world changing."

"It has long been known that in quantum mechanics these attitudes are equivalent: they predict exactly the same results for any experiment. Now, with the shadowing lemma, we find this same conclusion in chaotic systems," Hubbard summed up.

Implications for Biology?

If the "butterfly effect" is a striking symbol of our inability to predict the workings of the universe, Hubbard believes that another famous symbol of dynamical systems, the Mandelbrot set, suggests a way of thinking about two complex and vital questions: how a human being develops from a fertilized egg, and how evolution works.

He used a video to demonstrate the set, named after Benoit Mandelbrot, who recognized the importance of the property known as self-similarity in the study of natural objects, founding the field of fractal geometry (Box 3.3).

The largest view of the Mandelbrot set looks like a symmetric beetle or gingerbread man, with nodes or balls attached (Figure 3.3); looked at more closely, these balls or nodes turn out to be miniature copies of the "parent" Mandelbrot set, each in a different and complex setting (Figure 3.4). In between these nodes, a stunning array of geometrical shapes emerges: beautiful, sweeping, intricate curves and arcs and swirls and scrolls and curlicues, shapes that evoke marching elephants, prancing seahorses, slithering insects, and coiled snakes, lightning flashes and showers of stars and fireworks, all emanating from the depths of the screen.

For Hubbard, what is astounding about the Mandelbrot set is that its infinite complexity is set in motion by such a simple formula; the computer program to create the Mandelbrot set is only about 10 lines long. "If someone had come forth 40 years ago with a portfolio of a few hundred of these pictures and had just handed them out to the scientific community, what would have happened?" he asked. "They would presumably have been classified according to curlicues and this and that sticking out and so forth. No one would ever have dreamed that there was a 10-line program that created all of them."

BOX 3.3   JULIA SETS AND THE MANDELBROT SET

The Julia set and the Mandelbrot set, key mathematical objects of dynamical systems, are intimately related. The Julia set is a way to describe the sensitivity of a mathematical function to initial conditions: to show where it is chaotic. The Mandelbrot set summarizes the information of a great many Julia sets.

For there is not just one Julia set. Each quadratic polynomial ($z^2 +$ c, c being a constant), has its own Julia set—one for each value of c. The same is true for cubic polynomials, sines, exponentials, and other functions.

All Julia sets are in the complex plane. As we have seen, the Julia set for the simple quadratic polynomial $z^2 + c$, $c = 0$, is the unit circle: values of z outside the circle go to infinity on iteration; values inside go to 0.

But let c be some small number, and the Julia set becomes a fractal, McMullen said. In this case it is "a simple closed curve with dimension greater than 1." The curve is crinkly, "and this crinkliness persists at every scale." That is, if you were to magnify it, you would find that it was just as crinkly as before; it is self-similar. Other Julia sets—also fractals—form more intricate shapes yet, from swirls to clusters of tiny birds to crinkly lines suggestive of ice breaking on a pond.

The Mandelbrot set—some mathematicians prefer the term *quadratic bifurcation locus*—summarizes all the Julia sets for quadratic polynomials. Julia sets can be roughly divided into those that are connected and those that are not. (A set is *disconnected* if it is possible to divide it into two or more parts by drawing lines around the different parts.) The Mandelbrot set is formed of all the values of c for which the corresponding quadratic Julia set is connected.

Thus the Mandelbrot set is a prime example of a data-compressible system, that is, one created by a program containing far less information than it exhibits itself. It is, although complex, far from random, a random system being one that cannot be specified by a computer program or algorithm that contains less information than the system itself.

Hubbard suggested that the concept of data compressibility (sometimes known as low information density), a notion borrowed from computer science, could be useful in biology. "It is, I am almost sure, one of the really fundamental notions that we are all going to have to come to terms with."

"The real shock comes when you realize that biological objects

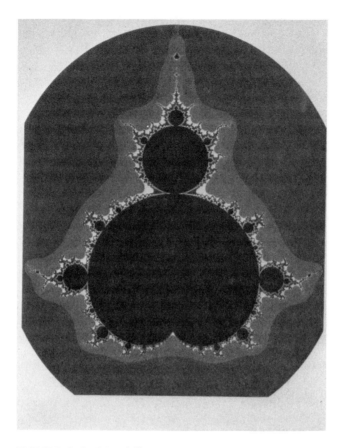

FIGURE 3.3 Mandelbrot set. Using computers as an experimental tool, mathematicians have found that simple rules can lead to surprisingly complicated structures. The computer program that creates the Mandelbrot set is only about 10 lines long, but the set is infinitely complex: the greater the magnification, the more detail one sees. (Courtesy of J.H. Hubbard.)

are specified by DNA, and that DNA is unbelievably short," he said. The "computer program" that creates a human being—the DNA in the genome—contains about 4 billion bits (units of information). This may sound big, but it produces far more than 100 trillion cells, and "each cell itself is not a simple thing." This suggests that an adult human being is far more data compressible—by a factor of 10,000, perhaps 100,000—than the Mandelbrot set, which itself is extraordinarily compressible.

FIGURE 3.4 Blowup of the Mandelbrot set. A small Mandelbrot set appears deep within the Mandelbrot set. (Courtesy of J.H. Hubbard.)

Hubbard sees the relatively tiny amount of information in the human gene as "a marvelous message of hope. . . . A large part of why I did not become a biologist was precisely because I felt that although clearly it was the most interesting subject in the world, there was almost no hope of ever understanding anything in any adequate sense. Now it seems at least feasible that we could understand it."

It also makes evolution more credible, he said: "Instead of thinking about the elaboration of the eye, you can think in terms of getting one bit every 10 years. Well, one bit every 10 years doesn't sound so unreasonable, whereas evolving an eye, that doesn't really sound possible."

THE ROLE OF COMPUTERS AND COMPUTER GRAPHICS

Since chaotic systems abound, why did scientists not see them before? A simple but not misleading answer is that they needed the right lens through which to examine the data. Hubbard recounted a conversation he had with Lars Ahlfors, the first recipient of the Fields Medal, the mathematical equivalent of the Nobel Prize. "Ahlfors is a great mathematician; no one could possibly doubt his stature. In fact, he proved many of the fundamental theorems that go into the subject. He tells me," Hubbard said,

> that as a young man, just when the memoirs of Fatou and Julia were being published, he was made to read them as—of course—you had to when you were a young man studying mathematics. That was what was going on. And further, in his opinion, at that moment in history and in his education they represented (to put it as he did) the pits of complex analysis. I am quoting exactly. He admits that he never understood what it was that they were trying to talk about until he saw the pictures that Mandelbrot and I were showing him around 1980. So, that is my answer to the question of the significance of the computer to mathematics. You cannot communicate the subject—even to the highest-level mathematician—without the pictures.

McMullen expressed a similar view: "The computer is a fantastic vehicle for transforming mathematical reality, for revealing something that was hitherto just suspended within the consciousness of a single person. I find it indispensable. I think many mathematicians do today."

Moreover, computer pictures of chaos and dynamical systems have succeeded—perhaps for the first time since Newton's ideas were argued in the salons of Paris—in interesting the public in mathematics. The pictures have been featured prominently in probably every major science magazine in the country and appear on tote bags, postcards, and teeshirts. Mandelbrot's 1975 book, *Les objets fractals: forme, hasard et dimension* (an English translation, *Fractals: Form, Chance and Dimension*, was published in 1977), which maintained that a great many shapes and forms in the world are self-similar, repeating their pattern at ever-diminishing scales, attracted widespread attention, and James Gleick's *Chaos: Making a New Science* was a surprise bestseller.

To some extent this attention is very welcome. Mathematicians have long deplored the public's unawareness that mathematics is an active field. Indeed, according to Krantz, the most likely explanation why no Nobel Prize exists for mathematics is not the well-known story that the mathematician Mittag-Leffler had an affair with Nobel's wife (Nobel had no wife) but simply that "Nobel was completely unaware that mathematicians exist or even if they exist that they do anything."

The mathematicians at the Frontiers symposium have tried to combat this ignorance. Hubbard, who believes strongly that research and education are inseparable, teaches dynamical systems to undergraduates at Cornell University and has written a textbook that uses computer programs to teach differential equations. Devaney has written not only a college-level text on dynamical systems, but also a much more accessible version, designed to entice high school students into learning about dynamical systems with personal computers. Thurston has developed a "math for poets" course at Princeton—"Geometry and the Imagination"—and following the symposium taught it to bright high school students in Minnesota.

"The fact that this mathematics is on the one hand so beautiful and on the other hand so accessible means that it is a tremendous boon for those of us in mathematics," Devaney said. "All of you in the natural sciences, the atmospheric sciences, astrophysicists, you all have these incredibly beautiful pictures that you can show your students. We in mathematics only had the quadratic formula and the Pythagorean theorem. . . . The big application of all of this big revolution will be in mathematics education, and God knows we can use it."

One of the scientists in the audience questioned this use of pictures, arguing that "the problem with science is not exciting people about science. . . . [It] is providing access. The reason that people are not interested in science is because they don't do science, and that is what has to be done, not showing pretty pictures." Devaney responded:

> You say that you have worked with elementary school students, and there is no need to excite them about science. That is probably true. That indicates that—as natural scientists—you have won the battle. In mathematics, however, this is far from the case. I run projects for high school students and high school teachers, and they are not excited about mathematics. They know nothing about the mathematical enterprise. They continually say to us, "What can you do with mathematics but teach?" There is no awareness of the mathematical enterprise or even of what a mathematician does.

But the popularity of the chaotic dynamical systems field, accompanied by what mathematicians at the session termed "hype," has also led to concerns and uneasiness. "I think chaos as a new science is given to a tremendous amount of exaggeration and overstatement," Devaney said. "People who make statements suggesting that chaos is the third great revolution in 20th-century science, following quantum mechanics and relativity, are quite easy targets for a comeback—

'Yes, I see. Something like the three great developments in 20th-century engineering: namely, the computer, the airplane, and the poptop aluminum can.'"

Some mathematicians, Krantz among them, feel that the public visibility of chaos is skewing funding decisions, diverting money from equally important, but less visible, mathematics. They also fear that standards of rigor and proof may be compromised. After all, the heart of mathematics is proof: theorems, not theories.

In his talk, Hubbard had related a telling conversation with physicist Mitchell Feigenbaum: "Feigenbaum told me that as far as he was concerned, when he thought he really understood how things were working, he was done. What really struck me is that as far as I was concerned, when I thought I really understood what was going on, that was where my work started. I mean once I really think I understand it, then I can sit down and try to prove it."

Part of the difficulty, Devaney suggested, is the very availability of the computer as an experimental tool, with which mathematicians can play, varying parameters and seeing what happens.

"With the computer now—finally—mathematics has its own essentially experimental component," Devaney told fellow scientists at the symposium. "You in biology and physics have had that for years and have had to wrestle with the questions of method and the like that arise. We mathematicians have never had the experimental component until the 1980s, and that is causing a few adjustment problems."

Krantz—who does not himself work in dynamical systems—has served as something of a lightning rod for this controversy following publication in *The Mathematical Intelligencer* of his critique of fractal geometry (Krantz, 1989). (He originally wrote the article as a review of two books on fractals for the *Bulletin of the American Mathematical Society*. The book reviews editor first accepted, and then rejected, the review. The article in *The Mathematical Intelligencer* was followed by a rebuttal by Mandelbrot.)

In that article, he said that the work of Douady, Hubbard, and other researchers on dynamical systems and iterative processes "is some of the very best mathematics being done today. But these mathematicians don't study fractals—they prove beautiful theorems." In contrast, he said, "fractal geometry has not solved any problems. It is not even clear that it has created any new ones."

At the symposium, Krantz said that mathematicians "have never had this kind of attention before," and questioned "if this is the kind of attention we want."

"It seems to me that mathematical method, scientific method, is something that has been hard won over a period of many hundreds, indeed thousands, of years. In mathematics, the only way that you know something is absolutely true is by coming up with a proof. In science, it seems to me the only way you know something is true is by the standard experimental method. You first become educated. You develop hypotheses. You very carefully design an experiment. You perform the experiment, and you analyze it and you try to draw some conclusions."

"Now, I believe absolutely in approximation, but successful approximation should have an error estimate and should lead to some conclusion. One does not typically see this in fractal geometry. Mandelbrot himself says that typically what one does is one draws pictures."

Indeed, Devaney remarked, "many of the lectures one hears are just a bunch of pretty pictures without the mathematical content. In my view, that is tremendously unfortunate because there is a tremendous amount of beautiful mathematics. Beyond that there is a tremendous amount of accessible mathematics."

Yet these concerns seem relatively minor compared to the insights that have been achieved. It may well distress mathematicians that claims for chaos have been exaggerated, that the field has inspired some less than rigorous work, and that the public may marvel at the superficial beauty of the pictures while ignoring the more profound, intrinsic beauty of their mathematical meaning, or the beauty of other, less easily illustrated, mathematics.

But this does not detract from the accomplishments of the field. The fundamental realization that nonlinear processes are sensitive to initial conditions—that one uses linear approximations at one's peril—has enormous ramifications for all of science. In providing that insight, and the tools to explore it further, dynamical systems is also helping to return mathematics to its historical position, which mathematician Morris Kline has described as "man's finest creation for the investigation of nature" (Kline, 1959, p. vii).

Historically, there was no clear division between physics and mathematics. "In every department of physical science there is only so much science . . . as there is mathematics," wrote Immanuel Kant (Kline, 1959, p. vii), and the great mathematicians of the past, such as Newton, Gauss, and Euler, frequently worked on real problems, celestial mechanics being a prime example. But early in this century the subjects diverged. At one extreme, some mathematicians affected a fastidious aversion to getting their hands dirty with real problems. Mathematicians "may be justified in rejoicing that there is one science at any rate, and that their own, whose very remoteness from ordinary human activities should keep it gentle and clean," wrote G.H. Hardy (1877–1947) in *A Mathematician's Apology* (Hardy, 1967, p. 121). For their part, physicists often assumed that once they had learned some rough-and-dirty techniques for using such mathematics as calculus and differential equations, mathematics had nothing further to teach them.

Over the past couple of decades, work in dynamical systems (and in other fields, such as gauge theory and string theory) has resulted in a rapprochement between science and mathematics and has blurred the once-clear line between pure and applied mathematics. Scientists are seeing that mathematics is not just a dead language useful in scientific computation: it is a tool for thinking and learning about the world. Mathematicians are seeing that it is exciting, not demeaning, when the abstract creations of their imagination turn out to be relevant to understanding the real world. They both stand to gain, as do we all.

BIBLIOGRAPHY

Hardy, G.H. 1967. A Mathematician's Apology. Cambridge University Press, Cambridge.

Hubbard, John H., and Beverly H. West. 1991. Differential Equations: A Dynamical Systems Approach. Part I: Ordinary Differential Equations. Springer-Verlag, New York.

Kline, Morris. 1959. Mathematics and the Physical World. Thomas Y. Crowell, New York.

Krantz, Steven. 1989. The Mathematical Intelligencer 11(4):12-16.

Lorenz, E.N. 1963. Deterministic non-periodic flows. Journal of Atmospheric Science 20:130-141.

Moser, Jürgen. 1973. Stable and Random Motions in Dynamical Systems. Princeton University Press, Princeton, N.J.

4

Looking Farther in Space and Time

For centuries by the practice of astronomy—first with the unaided human eye, then compass and sextant, finally with ever more powerful optical telescopes—humankind has undertaken to map the starlit skies, in hopes of better guiding navigation, constructing accurate calendars, and understanding natural cycles. Archaeological discoveries suggest that astronomy may be the oldest science, and Stonehenge one of the most prodigious, if not among the earliest, outdoor observatories. Employing a vast armamentarium of new imaging tools, the modern science of astrophysics begins with telescopic views of the visible sky. But astrophysicists are then able to use those observations to explore the physics of gravity, to observe the actions of fundamental particles, and to study other more exotic phenomena in space. From such explorations they develop theories about the origin, formation, structure, and future of the universe—the study of cosmology.

Guided by such theories, and using technology developed in the last several decades, astrophysicists probe phenomena beyond the range of vision. Beyond, that is, not only the farthest reach of optical telescopes (that are limited by the diameter of their mirrors and the sensitivity of their detecting devices) but also outside the relatively narrow range of the electromagnetic spectrum where visible-light rays reside, with wavelengths spanning from about 3000 to 9000 angstroms ($1 \text{ Å} = 10^{-8}$ cm). The modern astrophysicist harvests a much wider

NOTE: Addison Greenwood collaborated on this chapter with Marcia F. Bartusiak.

band of radiation from space, from radio signals with wavelengths meters in length to x rays and gamma rays, whose wavelengths are comparable to the size of an atomic nucleus (10^{-5} angstrom). And one of the most urgent and compelling questions in modern astrophysics concerns matter in space that apparently emits no radiation at all, and hence is referred to as dark matter.

J. Anthony Tyson from AT&T Bell Laboratories told the Frontiers symposium about his deep-space imaging project and the search for the dark matter that most astrophysicists now believe constitutes over 90 percent of all of the physical matter in the universe. If there really is 10 times as much invisible matter as that contained in luminous sources (and maybe even more), most theories about the evolution of the universe and its present structure and behavior will be greatly affected. Ever since the 1930s—soon after modern theories about the universe's evolution, now embraced as the Big Bang paradigm, were first proposed—scientists have been speculating about this "missing mass or missing light problem," explained Margaret Geller from the Harvard-Smithsonian Center for Astrophysics.

The Big Bang model has been generally accepted for about three decades as the reigning view of the creation of the universe. The whimsical name *Big Bang* was coined by British cosmologist Fred Hoyle (ironically, one of the proponents of the competing steady-state theory of the universe) during a series of British Broadcasting Corporation radio programs in the late 1940s. The Big Bang hypothesis suggests that the universe—all of space and matter and time itself—was created in a single instant, some 10 billion to 20 billion years ago. This event remains a subject of central interest in modern cosmology and astrophysics, both because its complexities are not fully understood and also because the world ever since—according to the credo of science and the dictates of natural law—may be understood as its sequel.

The event defies description, even in figurative language, but physicists have written an incredibly precise scenario of what probably occurred. There was not so much an explosion as a sudden expansion, away from an infinitely dense point that contained all of the matter of the universe. Both space and time, as currently conceived, did not exist before this moment but were created as the very fabric of the expansion. As space expanded at unimaginable speed, the matter it contained was carried along with it. Today's scientists can observe the resonant echos of this seminal event in many ways. The most significant consequence of the Big Bang is that nearly all objects in the universe are moving away from one another as the expansion continues, and the farther they are apart, the faster they are receding.

GRAVITY'S SHADOW: THE HUNT FOR OMEGA

Evidence for a dark, unknown matter permeating the universe is rooted in Newton's basic law of gravitation. In what may be the most influential science document ever written, the *Principia Mathematica*, the illustrious British scientist Isaac Newton in 1687 showed mathematically that there exists a force of attraction between any two bodies. Although Newton was working from observations made by the Danish astronomer Tycho Brahe and calculations earlier completed by Brahe's German assistant, Johannes Kepler, he extrapolated this work on celestial bodies to any two objects with mass, such as the apocryphal apple falling on his head. The attractive force between any two bodies, Newton demonstrated, increased as the product of their masses and decreased as the square of the distance between them. This revelation led to a systematic way of calculating the velocities and masses of objects hurtling through space.

Indeed, it was by applying Newton's formulae to certain observational data in the 1930s that Fritz Zwicky, a Swiss-born astronomer working at the California Institute of Technology, obtained astronomy's first inkling that dark matter was present in the universe. Concern about dark matter, noted Geller, "is not a new issue." By analyzing the velocities of the galaxies in the famous Coma cluster, Zwicky noticed that they were moving around within the cluster at a fairly rapid pace. Adding up all the light being emitted by the Coma galaxies, he realized that there was not enough visible, or luminous, matter to gravitationally bind the speeding galaxies to one another. He had to assume that some kind of dark, unseen matter pervaded the cluster to provide the additional gravitational glue. "Zwicky's dynamical estimates—based on Newton's laws which we think work everywhere in the universe—indicated that there was at least 10 times as much mass in the cluster of galaxies as he could account for from the amount of light he saw," said Geller. "The problem he posed we haven't yet solved, but we have made it grander by giving it a new name: Now we call it the dark matter problem." Over the last 15 years, astronomer Vera Rubin with the Carnegie Institution of Washington has brought the dark matter problem closer to home through her telescopic study of the rotations of dozens of spiral galaxies. The fast spins she is measuring, indicating rates higher than anyone expected, suggest that individual galaxies are enshrouded in dark haloes of matter as well; otherwise each galaxy would fly apart.

One of the reasons astrophysicists refer to the dark matter as a *problem* is related to the potential gravitational effect of all that dark matter on the fate of the cosmos, a question that arises when the equations of Einstein's theory of general relativity are applied to the universe at large. General relativity taught cosmologists that there is an intimate relationship between matter, gravity, and the curvature of space-time. Matter, said Einstein, causes space to warp and bend. Thus if there is not enough matter in the cosmos to exert the gravitational muscle needed to halt the expansion of space-time, then our universe will remain "open," destined to expand for all eternity. A mass-poor space curves out like a saddle whose edges go off to infinity, fated never to meet. The Big Bang thus would become the Big Chill. On the other hand, a higher density would provide enough gravity to lasso the speeding galaxies—slowing them down at first, then drawing them inward until space-time curls back up in a fiery Big Crunch. Here space-time is "closed," encompassing a finite volume and yet having no boundaries. With a very special amount of matter in the universe, what astrophysicists call a critical density, the universe would stand poised between open and closed. It would be a geometrically flat universe.

Scientists refer to this Scylla and Charybdis dilemma by the Greek letter Ω. Omega is the ratio of the universe's true density, the amount of mass per volume in the universe, to the critical density, the amount needed to achieve a flat universe and to just overcome the expansion. An Ω of 1 indicates a flat universe, greater than 1 a closed universe, and less than 1 an open universe. By counting up all the luminous matter visible in the heavens, astronomers arrive at a value for Ω of far less than 1; it ranges between 0.005 and 0.01. When astronomers take into account the dark matter measured dynamically, though, the amounts needed to keep galaxies and clusters stable, Ω increases to about 0.1 to 0.2. But theoretically, Ω may actually be much higher. In 1980 physicist Alan Guth, now with the Massachusetts Institute of Technology, introduced the idea that the universe, about 10^{-36} second after its birth, may have experienced a fleeting burst of hyperexpansion—an inflation—that pushed the universe's curvature to flatness, to the brink between open and closed. And with the geometry of the universe so intimately linked with its density, this suggests that there could be 100 times more matter in the cosmos than that currently viewed through telescopes.

THE TEXTURE OF THE UNIVERSE

Despite Guth's prediction, astronomical observations continue to suggest that the universe is open, without boundaries. As David Koo from the University of California, Santa Cruz, concluded, "If omega is less than one, the universe will not collapse. Presumably, we will die by ice rather than by fire." Yet astrophysicists look eagerly for hints of further amounts of dark matter by studying the various structures observed in the celestial sky. How the universe forged its present structures—galaxies grouping to form clusters, and the clusters themselves aggregating into superclusters—is significantly dependent on both the nature and amounts of dark matter filling the universe. The dark matter provides vital information, said Edmund Bertschinger of the Massachusetts Institute of Technology, "for the reigning cosmological paradigm of gravitational instability." Had there not been matter of a certain mass several hundred thousand years after the Big Bang, the universe would have been smooth rather than "clumpy," and the interactions between aggregations of particles necessary to form structures would not have taken place. S. George Djorgovski from the California Institute of Technology agreed that the "grand unifying theme of the discussion at the symposium—and for that matter in most of the cosmology today—is the origin, the formation, and the evolution of the large-scale structure of the universe," which, he said, "clearly must be dominated by the dark matter."

Geller and another guest at the symposium, Simon White from Cambridge University in Great Britain, have for years been involved in pioneering research to elucidate this large-scale structure. Until the 1980s, the best map of the universe at large was based on a vast catalog of galaxy counts, around 1 million, published in 1967 by Donald Shane and Carl Wirtanen from the Lick Observatory in California. They compiled their data from an extensive set of photographs taken of the entire celestial sky as seen from the Lick Observatory. Shane and Wirtanen spent 12 years counting each and every galaxy on the plates, and from their catalog P. James E. Peebles and his colleagues at Princeton University later derived the first major map of the universe, albeit a two-dimensional one. Peebles' 1-million-galaxy map was striking: The universe appeared to be filled with a network of knotlike galaxy clusters, filamentary superclusters, and vast regions devoid of galaxies. But was that only an illusion?

The raw input for Peebles' map did not include redshift data—data for establishing the third point of reference, the absolute distances of the various galaxies. How does one know, from a two-dimensional view alone, that the apparent shape of a group of stars (or galaxies) is not merely an accidental juxtaposition, the superposition of widely scattered points of light (along that particular line of sight) into a pattern discernible only because a human being is stationed at the apex? Without a parallax or triangulated view, or some other indicator of the depth of the picture being sketched, many unsupported suppositions were made, abetted perhaps by the inherent tendency of the human brain to impose a pattern on the heterogeneous input it receives from the eye.

The universe does display inhomogeneity; the question is how much. Soon after galaxies were discovered by Hubble in the 1920s, the world's leading extragalactic surveyors noticed that many of the galaxies gathered, in proximity and movement, into small groups and even larger clusters. The Milky Way and another large spiraling galaxy called Andromeda serve as gravitational anchors for about 20 other smaller galaxies that form an association called the Local Group, some 4 million light-years in width (a light-year being the distance light travels in a year, about 6 trillion miles). Other clusters are dramatically larger than ours, with over 1000 distinct galaxies moving about together. Furthermore, our Local Group is caught at the edge of an even larger assembly of galaxies, known as the Local Supercluster. Superclusters, which can span some 10^8 light-years across, constitute one of the largest structures discernible in astronomical surveys to date.

In mapping specific regions of the celestial sky, astronomers in the 1970s began to report that many galaxies and clusters appeared to be strung out along lengthy curved chains separated by vast regions of galaxyless space called voids. Hints of such structures also emerged when Marc Davis and John Huchra at the Harvard-Smithsonian Center for Astrophysics completed the first comprehensive redshift survey of the heavens in 1981, which suggested a "frothiness" to the universe's structure, a pattern that dramatically came into focus when Geller and Huchra extended the redshift survey, starting in 1985. Probing nearly two times farther into space than the first survey, the second effort has now pegged the locations of thousands of additional galaxies in a series of narrow wedges, each 6 degrees thick, that stretch across the celestial sky. Geller and her associates found, to their surprise, that galaxies were not linked with one another to form lacy threads, as previous evidence had been suggesting. Rather, galaxies appear to congregate along the surfaces of gigantic, nested bubbles, which Geller immediately likened to a "kitchen sink full of soapsuds." The huge voids that astronomers had been sighting were simply the interiors of these immense, sharply defined spherical shells of galaxies. Each bubble stretches several hundreds of millions of light-years across.

White and his collaborators, hunting constantly for clues to this bubbly large-scale structure, have developed a powerful computer simulation of its formation. This domain allows them to probe the nonequilibrium, nonlinear dynamics of gravitating systems. They have applied a great deal of theoretical work to the dark matter problem, helping to develop current ideas about the collapse of protogalaxies, how filaments and voids form in the large-scale distribution of clusters, and how galaxy collisions and mergers may have contributed to the evolution of structure in the present universe. They have used their numerical simulation technique to explore the types of dark matter that might give rise to such structures, such as neutrinos (stable, elementary particles with possibly a small rest mass, no charge, and an extreme tendency to avoid detection or interaction with matter) and other kinds of weakly interacting particles that have been hypothesized but not yet discovered. They link various explanations of dark-matter formation and distribution to the large-scale galaxy clustering that seems—as astrophysicists reach farther into deep space—to be ubiquitous at all scales.

But cosmologists look at the universe from many different perspectives, and many astrophysicists interested in galaxy formation and structure energetically search the skies in hopes of finding a galaxy at its initial stage of creation. Djorgovski goes so far as to call the search for a new galaxy aborning "the holy grail of modern cosmology. Though I suspect the situation is far more complicated, and I am not expecting *the* discovery of primeval galaxies." Rather, he predicted, "we will just learn slowly how they form," by uncovering pieces of the puzzle from various observations. One such effort he calls "paleontocosmology: since we can bring much more detail to our studies of nearby [and therefore older] galaxies than those far away, we can try to deduce from the systematics of their observed properties how they may have been created."

This ability to look back in time is possible because of the vast distances the radiation from the galaxies under observation must travel to reach earthbound observers. All electromagnetic radiation travels at a speed of approximately 300,000 kilometers per second (in a year about 9 trillion kilometers or 6 trillion miles, 1 light-year). Actually, the distances are more often described by astrophysicists with another measure, the parallax-second, or parsec, which equals 3.26 light-years. This is the distance at which a star would have a parallax equal to 1 second of arc on the sky. Thus 1 megaparsec is a convenient way of referring to 3.26 million light-years.

The laws of physics, as far as is known, limit the velocity of visible-light photons or any other electromagnetic radiation that is emitted in the universe. By the time that radiation has traveled a certain distance, e.g., 150 million light-years, into the range of our detection, it has taken 150 million years—as time is measured on Earth—to do so. Thus this radiation represents information about the state of its emitting source that far back in time. The limit of this view obviously is set at the distance light can have traveled in the time elapsed since the Big Bang. Astrophysicists refer to this limit as the observable universe, which provides a different prospect from any given point in the universe but emanates to a horizon from that point in all directions. Cosmologists study the stages of cosmic evolution by looking at the state of matter throughout the universe as the photons radiating from that matter arrive at Earth. As they look farther away, what they tend to see is hotter and is moving faster, both characteristics indicative of closer proximity to the initial event. The Big Bang paradigm allows cosmologists to establish points of reference in order to measure distances in deep space. As the universe ages, the horizon as viewed from Earth expands because light that was emitted earlier in time will have traveled the necessary greater distance to reach our view. At present, the horizon extends out to a distance of about 15 billion light-years. This represents roughly 50 times the average span of the voids situated between the richest superclusters recently observed. These superclusters are several tens of millions of light-years in extent, nearly 1000 times the size of our Milky Way galaxy. Our star, the Sun, is about 2 million light-years from its galactic neighbor, Andromeda. Nine orders of magnitude closer are the planets of its own solar system, and the Sun's light takes a mere 8 minutes to reach Earth.

Tyson's surveys excite astrophysicists because they provide data for galaxies more distant than those previously imaged, ergo, a window further back in time. Nonetheless, Jill Bechtold from the University of Arizona reminded symposium participants that "quasars are the most luminous objects we know, and we can see them at much greater distances than any of the galaxies in the pictures that Tony Tyson showed. Thus, you can use them to probe the distribution of the universe in retrospect," what astrophysicists call "lookback" time. *Quasar* is the term coined to indicate a quasistellar radio source, an object first recognized nearly 30 years ago. Quasars are thought to be the result of high-energy events in the nuclei of distant galaxies, and thus produce a luminance that can be seen much further than any individual star or galaxy.

Tyson and his fellow scientists took the symposium on a journey into deep space, showing pictures of some of the farthest galaxies ever seen. While these are the newest galaxies to be observed, their greater distance from Earth also means they are among the oldest ever seen, reckoned by cosmological, or lookback time. The astrophysicists described and explained the pictures and the myriad other data they are collecting with powerful new imaging tools, and also talked about the simulations and analyses they are performing on all of these with powerful new computing approaches. Along the way, the history of modern cosmology was sketched in: exploratory efforts to reach out into space, and back in time, toward the moment of creation some 10 billion to 20 billion years ago, and to build a set of cosmological theories that will explain how the universe got from there to here, and where—perhaps—it may be destined. In less than two decades, scientists have extended their probes to detect the radiation lying beyond the comparatively narrow spectrum of light, the only medium available to optical astronomers for centuries. The pace of theoretical and experimental advance is accelerating dramatically. The sense of more major discoveries is imminent.

THE BIG BANG PICTURE OF THE UNIVERSE

"The big bang," says Joseph Silk in his treatise *The Big Bang*, "is the modern version of creation" (Silk, 1989, p. 1). Silk, a professor of astronomy with the University of California at Berkeley, knits together most of the major issues in modern cosmology—the study of the large-scale structure and evolution of the universe—within the framework of the signal event most astrophysicists now believe gave birth to the universe around 15 billion years ago. Any such universal theory will rest on a fundamental cosmological principle, and Silk traces the roots of the Big Bang theory to Copernicus, who in 1543 placed the Sun at the center of the universe, displacing Earth from its long-held pivotal position. With this displacement of Earth from a preferred point in space came the recognition that our vantage point for regarding the universe is not central or in any way significant. Since observations show the arrangement of galaxies in space to vary little, regardless of direction or distance, scientists believe the universe is isotropic. This regularity is then imposed over two other basic observations: everything seems to be moving away from everything else, and the objects farthest removed from Earth are moving away proportionally faster. Taken together, these phenomena invite time into the universe and allow theorists to "run the film backwards" and deduce from their observations of current phenomena how the universe probably originated, in order to have evolved into what is manifest today. From this line of thinking, it follows that the universe began from a single point of infinitely dense mass, which underwent an unimaginable expansion some 10 billion to 20 billion years ago.

Although the astronomical data currently available favor a Big Bang cosmology, predictions based on the Big Bang model are continually examined. Perhaps the most convincing single test was accomplished in 1965, when radio astronomers Arno Penzias and Robert Wilson, using a highly sensitive receiver designed at their Bell Laboratories facility in New Jersey, detected what has come to be called the cosmic microwave background radiation. Spread uniformly across the celestial sky, this sea of microwaves represents the reverberating echo of the primordial explosion, a find that later garnered the Nobel Prize for Penzias and Wilson. Subsequent studies of this background radiation show it to be uniform to better than 1 part in 10,000. Such isotropy could arise only if the source were at the farthest reaches of the universe. Moreover, measurements of the radiation taken at different wavelengths, says Silk, indicate that it "originates from a state of perfect equilibrium: when matter and radiation are in equilibrium with one another, the temperatures of both must be identical" (Silk, 1989, p. 84).

Cosmologists call the point at the very beginning of time a singularity. Before it, classical gravitational physics can say or prove nothing, leaving all speculation to the metaphysicians. Big Bang theory encompasses a series of events that occurred thereafter, which conform to two continuing constraints: first, the laws of physics, which are believed to be universal, and second, data from observations that are continually probing farther in space, and therefore further back in time toward the event itself. This series of events cosmologists can "date," using either lookback time from the present or cosmic time. When measuring time forwards, singularity (the moment of creation) represents time zero on the cosmic calendar.

The events occurring during the first second have been hypothesized in great subatomic detail—and necessarily so—for at that point the temperature and pressure were greater than the forces that bind particles together into atoms. As time ensues, space expands, and matter thus thins out; therefore, temperature and pressure decrease. This "freezing" permits the forces that generate reactions between particles to accomplish their effects, in essence leading to the manufacture of a more complex universe. Complexity then becomes entwined with evolution, and eventually Earth and the universe reach their present state. Where the state of universal evolution has reached, cosmologists can only speculate. Since their observations rely on time machine data that only go backwards, they extrapolate according to the laws of physics into the future.

At the moment of creation, all four forces were indistinguishable; the most advanced theories in physics currently suggest that the forces were united as one ancestral force. But as the universe cooled and evolved, each force emerged on its own. At 10^{-43} second, gravity first separated, precipitating reactions between particles at unimaginably intense pressures and temperatures. At 10^{-36} second, the strong and the electroweak forces uncoupled, which in turn permitted the fundamental particles of nature, quarks and leptons, to take on their own identity. At 10^{-10} second, the four forces of the universe were at last distinct; as the weak force separated from the electromagnetic force, conditions deintensified to the point where they can be simulated by today's particle physicists exploring how matter behaves in giant particle accelerators. At 10^{-5} second, the quarks coalesced to form the elementary particles—called baryons—and the strong forces that bind the nucleus together came to dominate. By 1 second after singularity, the weak nuclear force allowed the decay of free neutrons into protons and the leptons, or light particles, the electrons, and neutrinos. Within 3 minutes, the protons and neutrons could join to form the light nuclei that even today make up the bulk of the universe, primarily hydrogen and helium. These nuclei joined electrons to form atoms within the next 1 million years, and only thereafter, through the evolution of particular stars, were these elements combined into the heavier elements described in the periodic table.

From the earliest moment of this scenario, the universe was expanding outward. The use of the term *Big Bang* unfortunately suggests the metaphor of an explosion on Earth, where mass particles hurtle outward from a central source, soon to be counteracted by the comparatively massive effects of the planet's gravity. In the universe according to Einstein, however, it is the fabric of space itself that is expanding—carrying with it all that matter and energy that emerged during creation's first microseconds, as described above. Gravity assumes a central role, according to Einstein's conception of the universe as an infinite fabric that evolves in a four-dimensional world called space-time. The mass that may be conceptualized as part of the fabric of space-time has a distorting effect on its shape, ultimately curving it. Einstein's ideas came to transform the face of science and the universe, but other astronomers and cosmologists, using general relativity as a touchstone, were the progenitors of the Big Bang conception that has come to define the modern view.

BASIC MEASUREMENTS IN SPACE

Beginning early in the 20th century, cosmology was revolutionized by astronomers making observations and developing theories that—in the context of first special and then general relativity—led in turn to more penetrating theories about the evolution of the universe. Such theories often required detection and measurement of stellar phenomena at greater and greater distances. These have been forthcoming, furthered by the development of larger optical telescopes to focus and harvest the light, spectrometers and more sensitive detectors to receive and discern it, and computer programs to analyze it. Improved measurements of the composition, distance, receding velocity, local motion, and luminosity of points of galactic light in the night sky provide the basis for fundamental ideas that underlie the Big Bang model and other conceptualizations about the nature of the universe. And radiation outside the optical spectrum that is detected on Earth and through probes sent into the solar system provides an alternative means of perceiving the universe.

With the recognition that the evolving universe may be the edifice constructed from blueprints conceived during the first few moments of creation has come the importance of particle physics as a context to test and explore conditions that are believed to have prevailed then. While many of those at the symposium refer to themselves still as astronomers (the original sky mappers), the term *astrophysicist* has become more common, perhaps for several reasons having to do with how central physics is to modern cosmology.

REDSHIFT PROVIDES THE THIRD DIMENSION

The development of the spectrograph in the 19th century brought about a significant change in the field of astronomy, for several reasons. First, since the starlight could be split into its constituent pieces (component wavelengths), the atomic materials generating it could at last be identified and a wealth of data about stellar processes collected. Second, spectrographic studies consist of two sorts of data. Bright emission lines radiate from the hot gases. Darker absorption lines are generated as the radiation passes through cooler gas on its journey outward from the core of a star. In the Sun, absorption occurs in the region known as the photosphere. Taken together, these lines reveal not only the composition of a source, but also the relative densities of its various atomic components. As an example, spectrographic studies of the Sun and other nearby stars indicate that 70 percent of the Sun is hydrogen and 28 percent helium, proportions of the two lightest elements that provide corroboration for the Big Bang model. Most stars reflect this proportion, and characteristic spectrographic signatures have also been developed for the major types of galaxies and clusters. But the third and most seminal contribution of spectroscopy to modern astrophysics was to provide a baseline for the fundamental measurement of cosmic distance, known as redshift.

The speed of light has been measured and is known. All radiation is essentially defined by its wavelength, and visible light is no exception. Light with a wavelength of around 4000 angstroms appears blue and that around 7000 angstroms appears red. Radiation with wavelengths just beyond these limits spills into (respectively) the ultraviolet and infrared regions, which our unaided eyes can no longer perceive. Starlight exhibits characteristic wavelengths because each type of atom in a star emits and absorbs specific wavelengths of radiation unique to it, creating a special spectral signature. Quantum physics permits scientists to postulate that the photons that constitute this light are traveling in waves from the star to an observer on Earth. If the source and observer are stationary, or are moving at equal velocities in the same direction, the waves will arrive at the receiving observer precisely as they were propagated from the source.

The Big Bang universe, however, does not meet this condition: the source of light is almost invariably moving away from the receiver. Thus, in any finite time when a given number of waves are emitted, the source will have moved farther from the observer, and that same number of waves will have to travel a greater distance to make the journey and arrive at their relatively receding destination than if the distance between source and observer were fixed. The wavelength expands, gets stretched, with the expanding universe. A longer wave is therefore perceived by the observer than was sent by the source. Again, since astrophysicists know the constituent nature of the light emitted by the stellar source as if it were at rest—from the unique pattern of spectral absorption and emission lines—they have input data for a crucial equation: Redshift = $z = (\lambda_{obs} - \lambda_{em})/\lambda_{em}$.

By knowing the wavelength the galaxy's light had when it was emitted (λ_{em}), they can measure the different wavelength it has when it is received, or observed (λ_{obs}). The difference permits them to calculate the speed at which the source of light is "running away"— relatively speaking—from the measuring instrument. The light from a receding object is increased in wavelength by the Doppler effect, in exactly the same way that the pitch of a receding ambulance siren is lowered. Any change in wavelength produces a different color, as perceived by the eye or by spectrometers that can discriminate to within a fraction of an angstrom. When the wave is longer, the color shift moves toward the red end of the optical spectrum. Redshift thus becomes a measurement of the speed of recession.

Most objects in the universe are moving away from Earth. When astrophysicists refer to greater redshift, they also imply an emitting source that is moving faster, is farther away, and therefore was younger and hotter when the radiation was emitted. Occasionally a stellar object will be moving in space toward Earth—the Andromeda galaxy is an example. The whole process then works in reverse, with photon waves "crowding" into a shorter distance. Such light is said to be "blueshifted." This picture "is true reasonably close to home," said Tyson. "When you get out to very, very large distances or large velocities, you have to make cosmological corrections." A redshift measurement of 1 extends about 8 billion years in look-back time, halfway back to creation. Quasars first emerged around 12 billion years ago, at a redshift of about 4. If we could see 13 billion years back to the time when astrophysicists think galaxy clustering began, the redshift (according to some theories) would be about 5. Galaxy formation probably extends over a wide range of redshifts, from less than 5 to more than 20. Theory suggests that redshift at the singularity would be infinite.

Thus early in the 20th century, astronomy—with the advent of spectrographs that allowed the measurement of redshifts—stood on the brink of a major breakthrough. The astronomer who took the dramatic step—which, boosted by relativity theory, quickly undermined then-current views of a static or stationary universe and provided the first strong observational evidence for the Big Bang paradigm—was Edwin Hubble. An accomplished boxer in college and a Rhodes scholar in classical law at Oxford University, who returned to his alma mater, the University of Chicago, to complete a doctorate in astronomy, Hubble eventually joined the staff of the Mount Wilson Observatory. His name has been memorialized in the partially-disabled space telescope launched in 1990, and also in what is probably the most central relationship in the theory of astrophysics.

Hubble's explorations with the 100-inch Hooker telescope on Mount Wilson were revolutionary. He demonstrated that many cloudlike nebulae in the celestial sky were in fact galaxies beyond the Milky Way, and that these galaxies contained millions of stars and were often grouped into clusters. From his work, the apparent size of the universe was dramatically expanded, and Hubble soon developed the use of redshift to indicate a galaxy's distance from us. He deduced that a redshift not only provided a measure of a galaxy's velocity but also indicated its distance. He used what cosmological distances were directly known—largely through the observations of Cepheid variable stars—and demonstrated that redshift was directly proportional to distance. That is, galaxies twice as far from his telescope were moving at twice the recessional speed.

The relationship between recessional velocity and distance is known as the Hubble constant, which measures the rate at which the universe is expanding. Tyson explained to the symposium that the expansion rate "was different in the past," because expansion slows down as the universe ages, due to deceleration by gravity. Agreeing on the value of the Hubble constant would allow cosmologists to "run the film backwards" and deduce the age of the universe as the reciprocal of the Hubble constant. Throughout this chapter, this book, and all of the scientific literature, however, one encounters a range rather than a number for the age of the universe, because competing groups attempting to determine the Hubble constant make different assumptions in analyzing their data. Depending on the method used to calculate the constant, the time since the Big Bang can generally vary between 10 billion and 20 billion years. Reducing this uncertainty in the measurement of the Hubble constant is one of the primary goals of cosmology.

As Koo pointed out, three of the largest questions in cosmology are tied up together in two measures, the Hubble constant and Ω. "How big is the universe? How old is the universe? And what is its destiny?" Values for the Hubble constant, said Koo, range from 50 to 100 kilometers per second per megaparsec. This puts the age of the universe between 10 billion and 20 billion years; bringing in various postulated values for Ω, the range can shift to between 6.5 billion and 13 billion years. Said Koo, "If you are a physicist or a theorist who prefers a large amount of dark matter so that Ω will be close to the critical amount," then the discovery by Tyson and others of huge numbers of galaxies in tiny patches of sky should be rather disturbing, because the theoretical calculations predict far fewer for this particular cosmology. Koo called this the "cosmological count conundrum." Conceding that "we, as astronomers, understand stars, perhaps, the best," Koo pointed out another possible anomaly, whereby the faintest galaxies are found to be unusually blue, and thus he calls this result "the cosmological color conundrum."

"We know that big, massive stars are usually in the very early stage of their evolution. They tend to be hot," explained Koo. And thus they appear very blue. Perhaps the faint galaxies are so blue because they have many more such massive stars in the distant past close to the epoch of their own birth. Resolving such puzzles, said Koo, will "push the limits of the age of the universe very tightly." If the factor of 2 in the uncertainty in the Hubble constant can be reduced, or the estimates of universal mass density refined, major theoretical breakthroughs could follow.

QUASAR STUDIES PROVIDE A MODEL

Radio surveys of objects (presumed to be galaxies) with intense radio-wavelength emissions (hypothesized as the radiation emitted by high-energy particles moving about in intense magnetic fields) were well established by the mid-1950s, and astronomers were beginning to use them to direct their optical telescopes, in hopes of further exploring their apparent sources. In 1963, while studying the spectrum of a radio source known as 3C 273, Maarten Schmidt at the California Institute of Technology was the first to identify quasars as extragalactic. Since that time, using a similar strategy, astronomers have identified thousands more. The name was coined to represent the phrase *quasi-stellar radio source*, indicating that, although these objects appeared starlike, they could not possibly be anything like a star. Their redshifts indicated they were far too distant to be resolvable as individual stars, nor could they be "plain vanilla" galaxies because of both the intensities (in many cases as great as that of hundreds of galaxies) and the dramatic fluctuations of their optical luminosities. Redshift measurements for 3C 273 indicate it is 1 billion to 2 billion light-years away; if so, its energy emission every second is equal to the energy our Sun generates over several hundreds of thousands of years.

Whatever they were, quasars suddenly provided a much deeper lens into the past. Explained Bechtold: "The high luminosities of quasars enable astronomers to study the universe at great distances and hence at epochs when the universe was about 20 percent of its current age. This was a period when many important events took place," not the least of which was "the rapid turn-on of quasars themselves." Quasar surveys of the sky indicate that the number of quasars per volume of space increases with the distance from the Milky Way, which suggests that whatever caused them was more common in the distant past, possibly during the era of galaxy formation. Astrophysicists have developed a consensus that quasars are the very active nuclei of galaxies, possibly the result of supermassive black holes at work. Black holes (as they are postulated) could cause the phenomenon astrophysicists detect with their spectrographs from quasars. If material were to collapse under gravity to a critical density, neither optical nor any other radiation could escape its grip. Einstein's theory of general relativity demonstrates this phenomenon with mathematical precision, and a parameter called the Schwarzschild radius provides its measure. For a mass the size of the Sun, the black hole will be compressed into a radius of 3 kilometers; black holes with larger masses increase their radii proportionally. Bechtold showed the symposium pictures of a quasar that would come from a black hole with between 10^7 and 10^9 solar masses.

"Material in the host galaxy that this black hole sits in falls into the black hole," Bechtold explained. Because it is approaching the object from its circular orbiting trajectory, "it cannot fall straight in," she added. "So, it sets up an accretion disk in the process of losing its angular momentum." The power of this reaction is incomparable, said Bechtold, and "is much more efficient in converting mass to energy than is the nuclear fusion that makes stars shine." As the material is drawn toward the center, swirling viscous forces develop and heat up the environment, "creating an incredible amount of energy that produces this continuum radiation we see" on our spectrographs. While quasars exhibit characteristic emission lines, other interesting spectral features arise in "little ensembles of clouds that live very close to the quasar," said Bechtold. "What is really most interesting about quasars is their absorption lines."

Because of the vast intergalactic distances traveled by quasar light, the chances of this light interacting with intervening material are great. Just as absorption lines are created as radiation from its core passes through the Sun's photosphere, so too does a quasar's emission spectrum become supplemented with absorption lines that display a smaller redshift, created when the radiation passes through "gas clouds along the line of sight between us and the quasar," explained Bechtold. "We are very fortunate that quasars have somehow been set up most of the way across the universe," she continued, "allowing us to study the properties of the interstellar medium in galaxies that are much too far away for us to be able to see the stars at all. Quasars allow us to probe galaxies at great distances and very large lookback times." As the signature quasar-emission-line radiation travels these vast distances, it almost inevitably encounters interstellar gas, as well as the gas presumed to reside in so-called haloes far beyond the optical borders of galaxies.

The most intriguing evidence from these probes, said Bechtold, is the distribution of the so-called Lyman alpha (Ly-α) absorption line, generated by hydrogen at 1215 angstroms. Analysis suggests that these data can provide vital clues, said Bechtold, to "the formation and rapid early evolution of galaxies, and the beginnings of the collapse of large structures such as clusters of galaxies and superclusters." Big Bang theory and numerous observations suggest that stars are the cauldron in which heavier elements are formed, after first hydrogen and then helium begin to aggregate through gravity. During such a stage, which may well be preceded by diffuse gas clouds of hydrogen throughout the region, no metals exist, and spectra from such objects are therefore devoid of any metal lines. Although Bechtold cautions about jumping to the conclusion that spectra without metal lines are definitive proof that no metals exist, she and other quasar specialists believe that "if there are, in fact, no metals, then these clouds probably predate the first stars. These may be the clouds of gas in the early universe with only hydrogen and helium that later formed stars. The stars then created elements like carbon, nitrogen, and oxygen through nuclear synthesis," but these "forests" of Lyman-alpha lines may provide the portrait of a protogalaxy that Djorgovski said astronomers eagerly seek (Figure 4.1).

While definitive confirmation of the full implications of the Lyman-alpha forest is in the future, Bechtold explained that the Lyman-alpha absorption lines possess an indisputable value for cosmology today. "Because they are so numerous, they can serve as tracers of large-scale structure very well, in fact, in some ways better than galaxies." Their distribution throughout the universe is "much more uniform than galaxies in the present day," and thus they provide a possible precursor or link in the causal chain that might explain the

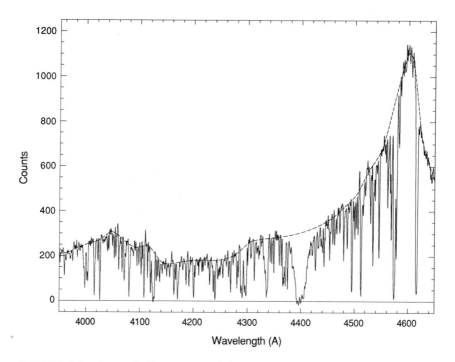

FIGURE 4.1 An optical spectrum of the quasar 0913+072 obtained with the Multiple Mirror Telescope on Mt. Hopkins, Arizona, a joint facility of the Smithsonian Institution and the University of Arizona. The dashed line indicates the quasar continuum; the deep absorption lines originate in galaxies along the line of sight, or intergalactic Lyman-alpha forest clouds. (Courtesy of J. Bechtold.)

present structure. "They don't cluster. They don't have voids," pointed out Bechtold, which "supports the idea that they are protogalactic or pregalactic clouds that have collapsed to galaxies and later the galaxies collapsed to form clusters of galaxies." Thus quasars provide penetrating probes into the past because of their distance, which, but for their intense luminosity, would preclude optical detection. This fact, together with the phenomenon of a distant source of radiation picking up information about intermediate points along the path to Earth, provides a parallel with another recent and dramatic development in astrophysics.

DARK MATTER SEEN THROUGH THE LENS OF GRAVITY

Revitalizing a method that is very old, Tyson and his colleagues have made use of the phenomenon of gravitational lensing, based on the theory of general relativity conceived by Einstein. Quantum mechanics considers light not only as a wave with a characteristic length, but alternatively as a particle called a photon. Einstein's special theory of relativity allows one to consider an effective but tiny mass for the photon of $m = E/c^2$, where E is its energy and c is the speed of light. If a photon of light from some far galaxy enters the gravitational field of a closer object (not unlike a rocket or satellite sailing past a planet), the gravitational force between the object and the photon will pull the photon off course, deflecting it at an angle inward, where it will continue on a new, straight-line trajectory after it has escaped the gravitational field. If that subsequent straight line brings it within the purview of a telescope, scientists observing it can apply the formulas proposed by Einstein, as well as the simple geometry of Euclid. Knowing the angle of arrival of the photon relative to the location of the light-emitting source, as well as the distance to the mass, one can calculate the size of the mass responsible for altering the photon's path.

In 1915 Einstein had predicted just such a deflection of light by the Sun's gravitational field. It was twice the deflection predicted by classical Newtonian analysis. When England's Sir Arthur Eddington confirmed Einstein's prediction during a solar eclipse in 1919, general relativity won wide acceptance. Einstein refined the analysis years later, applying it to stars and predicting the possibility of what came to be called an Einstein ring, a gravitational lensing effect that would occur when a far-off star happened to line up right behind a star closer in. But Einstein dismissed such an alignment as being too improbable to be of practical interest. Zwicky, exhibiting his well-known prescience (he predicted the existence of neutron stars), went on to speculate that nearby galaxies, with masses 100 billion times greater than a star, would also act as gravitational lenses, splitting the light of more distant objects into multiple images.

Tyson picked up the story: "The first lensed object was discovered in 1979. It was a quasar seen twice—two images, nearby, of the same quasar—a sort of cosmic mirage." But the confirmation of Einstein, as Zwicky had predicted, was to vindicate their potential value as a sort of telescope of distant objects, a further source of information about the average large-scale properties of the universe, as well as an indicator of the presence of inhomogeneities in the universe, particularly those arising from dark matter. Tyson said that one of the reasons they seemed to offer a glimpse into the dark matter question was expressed in a seminal paper by Harvard University's William Press and James Gunn of Princeton that ruled out a universe closed by dark matter purely in the form of black holes. The technique was still considered limited, however, to the chance conjunction of a foreground gravitational lens with a distant quasar. In 1988, Edwin Turner of Princeton wrote that "only quasars are typically far enough away to have a significant chance of being aligned with a foreground object. Even among quasars, lens systems are rare; roughly 2,000 quasars had been catalogued before the first chance discovery of a lensed one in 1979" (Turner, 1988, p. 54). True enough, but if chance favors the prepared mind, Tyson's 10 years and more of deep-space imaging experience was perhaps the context for a new idea.

Discovering the Blue "Curtain"

Using the 4-meter telescopes situated on Kitt Peak in Arizona and at the Cerro Tololo Observatory in Chile, Tyson was trying to increase the depth of the surveys he was making into the sky. It was the distance of quasars that made them such likely candidates for lensing: the greater the intervening space, the greater the likelihood of fairly rare, near-syzygial alignment. Tyson was not looking exclusively for quasars, but was only trying to push back the edge of the telescopic horizon. Typically, existing earthbound telescopes using state-of-the-art photographic plates in the early 1970s were capturing useful galactic images out to about 2 billion to 3 billion light-years.

Tyson explained the brightness scale to his fellow scientists, alluding to the basic physics that intensity decreases as the square of the distance. "Magnitude is the astronomer's negative logarithmic measure of intensity. With very, very good eyesight, on a clear night, you can see sixth magnitude with your eye." As the magnitude decreases in number, the intensity of the image increases logarithmically—the brightest image we see is the Sun, at a magnitude of around –26. Though faint, the images Tyson was producing photographically registered objects of magnitude 24, at a time when the faintest objects generally thought to be of any use were much brighter, at about magnitude 20. Djorgovski supplied an analogy: "Twenty-fourth magnitude is almost exactly what a 100-watt light bulb would look like from a million kilometers away or an outhouse light bulb on the moon."

Although the images Tyson and others have produced photographically at this magnitude were extremely faint, they still revealed 10,000 dim galaxies on each plate covering half a degree on the sky. Astronomers for the last five decades have been straining their eyes on such photographs, and Tyson was doing the same when he began to realize that "my data were useless." In 1977, he had been counting these faint smudges for a while, and noticed that "as the week went on, I was counting more and more galaxies on each plate. I realized that I was simply getting more skillful at picking them out." Tyson fully realized the problem—the human element—and began to look for a more objective method of counting galaxies.

What he found reflects the impact of technology on modern astrophysics. He sought out John Jarvis, a computer expert in image processing, and together they and Frank Valdes developed a software program called FOCAS (faint object classification and analysis system). With it, the computer discriminates and classifies these faint smudges, producing a catalog from their size and shape that separates stars from galaxies, and catalogs a brightness, shape, size, and orientation for each galaxy. Now that he had established a scientifically reliable method for analyzing the data, he needed to image more distant galaxies (in order to use them as a tool in dark matter tests by gravitational lensing). In this pursuit he became one of the first astronomers to capitalize on a new technology that has revolutionized the field of faint imaging—charge-coupled devices, or CCDs, which act like electronic photographic plates. Even though many telescopes by now have been equipped with them, the usefulness of CCDs for statistical astronomy has been limited by their small size. Tyson and his collaborators are building a large-area CCD mosaic camera, which he showed to the scientists at the symposium, that will increase the efficiency. CCDs have dramatically increased the horizon for faint imaging because "they have about 100 times the quantum efficiency of photographic plates and are also extremely stable. Both of these features, it turns out," said Tyson, "are necessary for doing ultradeep imaging."

What happened next, in 1984, was reminiscent of Galileo's experience in the early 17th century. At that time Galileo had remarked that with his rudimentary telescope he could see "stars, which escape the unaided sight, so numerous as to be beyond belief" (Kristian and Blouke, 1982). Said Tyson, "Pat Seitzer and I chose a random region of the sky to do some deep imaging in. Based on what we knew at the time about galaxy evolution, and by going several magnitudes fainter with the aid of the CCDs, we expected to uncover perhaps 30 galaxies in this field." Previous photographic technology had shown none in such a field. Their region was "a couple of arc minutes across, about 1 percent of the area of the moon," explained Tyson, for comparison. But instead of the expected 30, their CCD exposure showed 300 faint blue objects, which FOCAS and other computer treatments helped to confirm were galaxies. They continued to conduct surveys of other random regions of about the same size and came up with about 10,000 faint galaxy images. The bottom line, based on reasonable extrapolation, was the discovery of 20 billion new galaxies over the entire celestial sky. In effect, these pictures showed a new "curtain" of galaxies, largely blue because of their early stage of evolution. There still was insufficient light to measure their redshift, but they clearly formed a background curtain far beyond the galaxies that could be measured.

Gravity's Role as a Lens and a Cosmological Force

Tyson and his colleagues were energetically conducting their surveys of this blue "curtain" even as other astronomers were discovering faint blue arcs, thought to be the result of gravitational lensing. The blue curtain is a much more efficient probe of foreground dark matter because the curtain is far enough away to enhance the chances of foreground lensing and rich enough to provide a canvas—rather than a point of light coming from a quasar—to observe the distortion. After subtracting the gravitational effects of those luminous galaxies in the foreground that they could observe, Tyson's team could nonetheless still observe fairly dramatic distortion in the blue curtain galaxy images. They realized they had taken an image—albeit an indirect one—of the dark, nonluminous matter in the foreground cluster of galaxies (Figure 4.2). With quasar studies, the foreground object was often a very specific galaxy or star, and the light was emanating from the discrete quasar source. Tyson's system works by the same principles of gravity, but instead of an identifiable body that one can see, the only evidence of the foreground object is the gravitational pull it exerts on the path of the distant light and the resulting tell-tale distortion of the distant galaxy's image (Figure 4.3).

By enhancing FOCAS and developing other programs to simulate

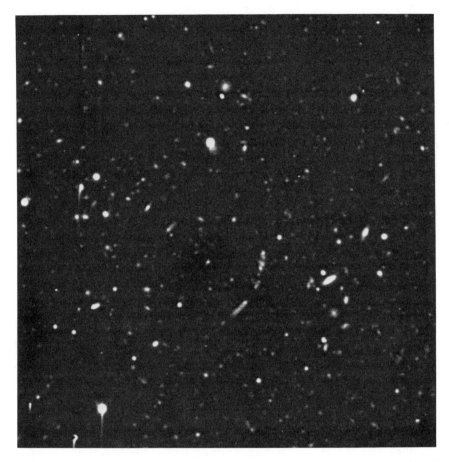

FIGURE 4.2 The gravitational effect of dark matter in a foreground cluster of galaxies. The images of the foreground galaxies have been eliminated, revealing faint arcs. The clump of dark matter is 4 billion light-years away. (Courtesy of J.A. Tyson.)

the possible shape and composition of dark matter, Tyson has been able to use his blue curtain as a vast new experimental database. Characteristic distortions in the previously too faint images have become observable as the deep CCD optical imaging surveys and analyses continue. A portrait of dark matter is emerging that allows other theories about the evolution of the universe to be tested—galaxy cluster formation, for one. "We have found that there is a morphological similarity between the distribution of the bright cluster galaxies and the distribution of the dark matter in the cluster, but the dark matter is distributed relatively uniformly on the scale of an individual galaxy in the cluster," said Tyson. Their lens technique is better at determining the mass within a radius, to within about 10 percent accuracy, than the size of the inferred object. "Less accurate," added Tyson, "is the determination of the core size of this distribution. That is a little bit slippery, and for better determination of that we simply have to have more galaxies in the background. But

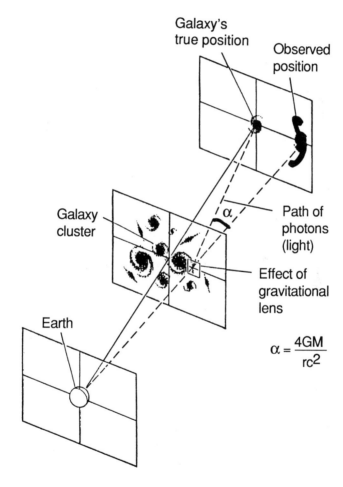

FIGURE 4.3 A diagram showing the bending of a light path by the mass in a gravitational lens and the apparent change in the position of the source and the resulting distortion of its image. (Courtesy of J.A. Tyson.)

since we are already going to around 30th magnitude or so in some of these fields, that is going to be hard to do."

"A very interesting question that is unanswered currently because of the finite size of our CCD imagers is how this dark matter really falls off with radius. Is the dark matter pattern that of an isothermal distribution of particles bound in their own gravitational potential well, in which case it would go on forever?" Tyson wondered. This is, in part, the impetus for Tyson to develop a larger CCD device.

Results from Tyson's work may aid astronomers in their quest to understand how galaxies formed. In order to create the lumps that eventually developed into galaxies, most theories of galaxy formation start with the premise that the Big Bang somehow sent a series of waves rippling through the newly born sea of particles, both large and small fluctuations in the density of gas. According to one popular theory of galaxy formation, small knots of matter, pushed and squeezed by those ripples, would be the first to coalesce. These lumps, once they evolve into separate galaxies, would then gravitationally gather into clusters and later superclusters as the universe evolves. This process has been tagged, appropriately enough, the "bottom-up" model.

Conversely, a competing theory of galaxy formation, known as the "top-down" model, essentially reverses the scale of collapse, with the largest, supercluster-sized structures collapsing first and most rapidly along the shorter axis, producing what astrophysicists call a pancake structure, which would then fragment and produce cellular and filamentary structures.

Bertschinger explained that gravitational instability causes "different rates of expansion depending on the local gravitational field strength. Gravity retards expansion more in regions of above-average density, so that they become even more dense relative to their surroundings. The opposite occurs in low-density regions."

Following earlier computer models simulating the evolution of a universe filled with cold dark matter, especially a model pioneered by Simon White and his associates in the 1980s, Bertschinger and his colleagues have explored how dark matter may cluster and how this development may parallel or diverge from galaxy clustering: "The gravitational field can be computed by a variety of techniques from Poisson's equation, and then each particle is advanced in time according to Newtonian physics." The complexity comes in when they try to capture the effects of the explosive microsecond of Guth's inflation era. For this they need to model the quantum mechanical fluctuations—essentially the noise—hypothesized to extend its impact through time by the medium of acoustic waves. In their model, said Bertschinger, "most of the dark matter does end up in lumps associated with galaxies. The lumps are something like 1 million light-years" in extent, but it is not conclusive that they correspond to the haloes found beyond the luminous edges of galaxies. "It is plausible," he continued, "that luminous galaxies evolve in the center of these lumps," but to say so more definitively would involve simulating the dynamics of the stellar gases, which was beyond the scope of their model. But dark matter does seem to cluster, and in a manner similar to galaxies. Most of it is found within 1 million light-years of a galaxy. These conclusions are tentative, in that they emerged largely from simulations, but they are consistent with many of the observed data, including the limits found with respect to the anisotropy of the cosmic microwave background.

THE FORMATION OF STARS

Djorgovski is another astrophysicist, like Bechtold and Tyson, whose studies about one phenomenon—the formation and evolution of galaxies—often touch on larger cosmological questions. On the assumption that dark matter constitutes 90 percent or more of the mass in the universe, "large-scale structure is clearly dominated" by it, he said. But galaxies are the components of this large-scale structure, and their structure "is a much more complicated business." Astronomers approach the subject in two ways, he said. Looking at large redshifts, they hope to "come upon a primeval galaxy that will hopefully be a normal example undergoing its first major burst of star formation." Another approach, called paleontocosmology, looks at nearby, older galaxies that can be studied in some detail. Systematic properties may yield scaling laws and correlations that in turn allow strong inferences to be made about the early life and formation of galaxies (Figure 4.4). "Galaxies are interesting per se," said Djorgovski, but better understanding their evolution could provide astrophysicists "tools to probe the global geometry, or the large-scale velocity field."

Galaxy formation can be seen as two distinct types of events, said Djorgovski: assembling the mass (which is the business of the dark matter and gravity) and "converting the primordial [hydrogen] gas, once assembled, into stars that shine." The energy for this second step comes from two sources, he explained. "First of all, when you look at galaxies, you find out that they are about 1000 times denser than the large-scale structure surrounding them. That tells you that they must have collapsed out of their surrounding density field by

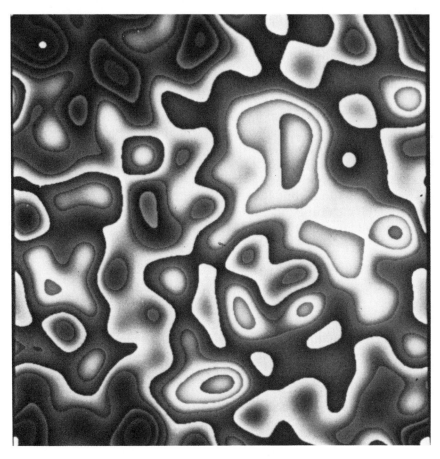

FIGURE 4.4 Computer simulation of a density field in the early universe. Shades of gray indicate different densities in a fairly arbitrary way. (Courtesy of S.G. Djorgovski.)

about a factor of 10, and that involves dissipation of energy. When you add up how much that is, it comes to about 10^{59} ergs per galaxy." Once formed, stars engage in nuclear fission to make the helium and heavier metals that come to compose their cores, and they generate about 100 times more energy, 10^{61} ergs per galaxy.

These events are thought to happen in the era around a redshift of 2, perhaps a little greater, "but we don't know exactly where or when," said Djorgovski, suggesting that the process probably evolves over a considerable period of time. But it will be the emission lines in its spectra that confirm a star in the process of forming, should one be found. Djorgovski showed the scientists some pictures of one candidate "which has the romantic name of 3C 326.1, at a redshift of nearly 2, two-thirds of the way to the Big Bang." Whether this is a primeval galaxy remains a subject for debate, as is the case with many such candidates, because it is associated with intense radio emissions and its signals may be confounded by an active nucleus it contains. "We know about a dozen objects of this sort," said Djorgovski, "but what you really want to find are just ordinary galaxies forming with large redshifts. And that nobody has found so far."

BIBLIOGRAPHY

Flamsteed, Sam. 1991. Probing the edge of the universe. Discover 12(7):40-47.

Guth, Alan H., and Paul J. Steinhardt. 1984. The inflationary universe. Scientific American 250(May):116-128.

Kristian, Jerome, and Morley Blouke. 1982. Charge-coupled devices in astronomy. Scientific American 247(October):66-74.

Silk, Joseph. 1989. The Big Bang. Freeman, New York.

Turner, Edwin. 1988. Gravitational lenses. Scientific American 259(July):54-60.

RECOMMENDED READING

Harrison, Edward 1981. Cosmology: The Science of the Universe. Cambridge University Press, Cambridge.

5

Gene Control: Transcription
Factors and Mechanisms

Since the elucidation of the double-helix structure of deoxyribonucleic acid (DNA) in 1953, biologists have been racing to understand the details of the science of genetics. The deeper they penetrate into the workings of the DNA process, however, the more complexity emerges, challenging the early optimism that characterizing the structural mechanisms would reveal the entire picture. It now appears likely that life within an organism unfolds as a dynamic process, guided by the DNA program to be sure, yet not subject to clockwork predictability. One of the most intriguing questions involves the very first step in the process, how the DNA itself delivers its information to the organism. At the Frontiers symposium, a handful of leading genetic scientists talked about their research on transcription—the crucial first stage in which RNA molecules are formed to deliver the DNA's instructions to a cell's ribosomal protein production factories. The discussion was grounded in an overview by Robert Tjian, who leads a team of researchers at the Howard Hughes Medical Institute and teaches in the Department of Molecular and Cell Biology at the University of California's Berkeley campus.

Eric Lander of the Whitehead Institute at the Massachusetts Institute of Technology organized the session "to give a coordinated picture of gene control in its many different manifestations, both the different biological problems to which it applies and the different methods people use for understanding it." The goal was to try to provide the "nonbiologists at the symposium with a sense of how the genome knows where its genes are and how it expresses those genes."

Possibly no other scientific discovery in the second half of the 20th century has had the impact on science and culture that elucidation of DNA's structure and function has had. The field of molecular biology has exploded into the forefront of the life sciences, and as its practitioners rapidly develop applications from these insights, new horizons appear continuously. The working elements of genetics, called genes, can now be duplicated and manufactured, and then reintroduced into living organisms, which generally accept them and follow their new instructions. Recombinant DNA technology and gene therapy promise to change not only our view of medicine, but also society's fundamental sense of control over its biological fate, perhaps even its evolution.

HOW DNA WORKS

Despite the significance of modern genetics, many of its fundamentals are still not widely understood. A summary of what has been learned about DNA might serve as a useful introduction to the discussion on transcription and gene expression:

• The heritable genetic information for all life comes in the form of a molecule called DNA. A full set of a plant's or an animal's DNA is located inside the nucleus of every cell in the organism. Intrinsic to the structure of the DNA molecule are very long strings composed of so-called base pairs, of which there are four types. A gene is a segment of this string that has a particular sequence of the four base pairs, giving it a unique character. Genes are linked one after another, and the string of DNA is carried on complex structures called chromosomes, of which there are 23 pairs in humans. Researchers put the number of discrete genes in humans at about 100,000. To clarify the concept of DNA, Douglas Hanahan from the University of California, San Francisco, invoked the metaphor of a magnetic tape, "which looks the same throughout, but has within it (or can have) discrete songs composed of information." A gene can thus be likened to a particular song.

• The general outline of this picture was known by the early 1950s, but even the electron microscope had not revealed exactly how the DNA molecule was structured. When British biophysicist Francis Crick and American molecular biologist James Watson first proposed the double-helix structure for DNA, a thunderclap echoed throughout molecular biology and biochemistry. Much more than just a classification of structure, it was a revelation whose implications opened up a vast area of exploration. Why so momentous? Because that structure facilitates DNA's role and function to such an extent that the whole process of decoding and eventually altering the basic genetic information was suddenly glimpsed by drawing the curtain back on what has come to be known as the alphabet of life.

• The structure of DNA was at once realized to be dramatically suggestive of how the molecule actually functions to store and deliver coded information. By weak chemical bonding between complementary bases—adenine with thymine and cytosine with guanine, and each pair vice versa—the hereditary store of information in all life forms takes shape as a coded sequence of simple signals. The signals are arranged in the double-helix structure discovered by Watson and Crick. Picture two strands of rope side by side, each with a string of chemical bases along its length (Figure 5.1). When a base on the first rope is adenine (A), the base opposite it on the other rope

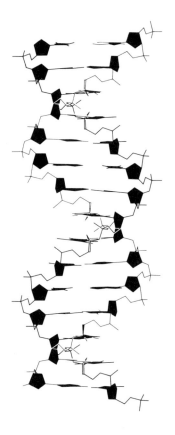

FIGURE 5.1 Skeletal model of double-helical DNA. The structure repeats at intervals of 34 angstroms, which corresponds to 10 residues on each chain. (From p. 77 in BIO-CHEMISTRY 3rd edition, by Lubert Stryer. Copyright © 1975, 1981, 1988 by Lubert Stryer. Reprinted by permission from W.H. Freeman and Company.)

will be thymine (T). Also conversely, if thymine appears on one strand, adenine will be found opposite on the other strand. The same logic applies to analogous pairings with cytosine (C) and guanine (G). These base pairs present the horizontal connection, as it were, by their affinity for a weak chemical bond with their complementary partner on the opposite strand. But along the vertical axis (the rope's length), any of the four bases may appear next. Thus the rope—call it a single strand, either the sense strand or the antisense strand—of DNA can have virtually any sequence of A, C, G, and T. The other strand will necessarily have the complementary sequence. The *code* is simply the sequence of base pairs, usually approached by looking at one of the strands only.

• In their quest to explain the complexity of life, scientists next turned to deciphering the code. Once it was realized that the four nucleotide bases were the basic letters of the genetic alphabet, the question became, How do they form the words? The answer was known within a decade: the 64 possible combinations of any given three of them—referred to as a triplet—taken as they are encountered strung along one strand of DNA, each delivered an instruction to "make an amino acid."

• Only 20 amino acids have been found in plant and animal cells. Fitting the 64 "word commands" to the 20 outcomes showed that a number of the amino acids could be commanded by more than one three-letter "word sequence," or nucleotide triplet, known as a codon (Figure 5.2). The explanation remains an interesting question, and so far the best guess seems to be the redundancy-as-error-protection theory: that for certain amino acids, codons that can be mistaken in a "typographical mistranslation" will not so readily produce a read-out error, because the same result is called for by several codons.

• The codons serve, said Hanahan "to transmit the protein coding information from the site of DNA storage, the cell's nucleus, to the site of protein synthesis, the cytoplasm. The vehicle for the transmission of information is RNA. DNA, the master copy of the code, remains behind in a cell's nucleus. RNA, a molecule whose structure is chemically very similar to DNA's, serves as a template for the information and carries it outside the cell's nucleus into the cytoplasm, where it is used to manufacture a given sequence of proteins.

Once the messenger transcript is made, its translation eventually results in the production (polymerization) of a series of amino acids that are strung together with peptide bonds into long, linear chains that in turn fold into interesting, often globular molecular shapes due to weak chemical affinities between and among various amino acids.

A

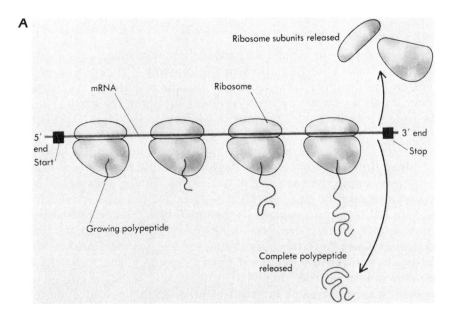

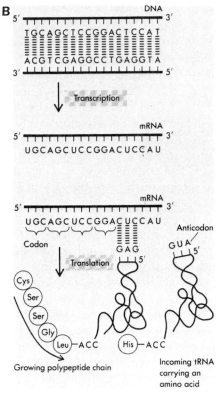

FIGURE 5.2 (A) Diagram of a polyribosome. Each ribosome attaches at a start signal at the 5' end of a messenger RNA (mRNA) chain and synthesizes a polypeptide as it proceeds along the molecule. Several ribosomes may be attached to one mRNA molecule at one time; the entire assembly is called a polyribosome. (B) Transcription and translation. The nucleotides of mRNA are assembled to form a complementary copy of one strand of DNA. Each group of three is a codon that is complementary to a group of three nucleotides in the anticodon region of a specific transfer (tRNA) molecule. When base pairing occurs, an amino acid carried at the other end of the tRNA molecule is added to the growing protein chain. (Reprinted with permission from Watson et al., 1987, p. 84. Copyright © 1987 by The Benjamin/Cummings Publishing Company, Inc.)

These steps are chemically interesting, but the mystery that compelled the scientists at the Frontiers symposium surrounds the initial transcript that is created by a species of ribonucleic acid called messenger RNA (mRNA). Tjian's overview, "Gene Regulation in Animal Cells: Transcription Factors and Mechanisms," touched on much of the above background and presented some of the basic issues scientists are exploring as they probe the mRNA process. His colleagues in the session on gene regulation each described intriguing findings based on their studies of regulation in various organisms: Arnold Berk, from the University of California, Los Angeles, in viruses; Kevin Struhl, from the Harvard Medical School, in yeast; Ruth Lehmann, from the Whitehead Institute, in the fruit fly; and Hanahan, in mice. They explained their work to the symposium and suggested how its implications may help to clarify human genetics and fill in the larger picture of how life operates.

THE ROLE OF DNA

Noncoding DNA—Subtlety, Punctuation, or Just Plain Junk?

Scientists cannot say for certain whether the majority of noncoding genes that do not seem to say simply "make this string of amino acids," are saying anything at all. Tjian has heard a lot of speculation on this question: "Most people in the field agree that only a very small percentage of the human genome is actually coding for . . . the proteins that actually do all the hard work. But far be it for me to say that all that intervening sequence is entirely unimportant. The fact is we don't know what they do." He notes an interesting phenomenon among scientists, observing that while "some people call it junk," others like himself "would rather say 'I don't know.'" Rather than dismissing the significance of this undeciphered DNA, he and others mentally classify it as punctuation that is modifying and influencing the transcript. "There is clearly a lot of punctuation going on, yet still the question arises: Why do amphibians have so much more DNA than we do? Why does the simple lily have so much DNA, while the human—clearly just as complicated in terms of metabolic processes—doesn't seem to need it? Actually, a lot of people wonder about whether those sequences are perhaps there for more subtle differences—differences between you and me that at our present stage of sophistication may be too difficult to discern."

Eric Lander, responding to the curiosity about excess or junk genes, pointed out that the question often posed is, If it is not useful, why is it there? He continued, "From an evolutionary point of view, of course, the relevant question is exactly the reverse: How would you get rid of it? It takes work by way of natural selection to get rid of things, and if it is not a problem, why would you junk it? That is really the way life is probably looking at it." Vestigial traits are not uncommon at the higher rungs of the evolutionary ladder, pointed out Lander, whereas "viruses, for example, are under much greater pressure to compete, and do so in part by replicating their DNA efficiently."

However, the astounding intricacy, precision, and timing of the biological machinery in our cells would seem to suggest to Tjian and others that the nucleotide base sequences in between clearly demarcated coding genes do have a vital function. Or more likely, a number of functions. Now that the questions being posed by scientists mapping the genome are starting to become more refined and subtle, the very definition of a gene is starting to wobble. It is often convenient to conceptualize genes as a string of discrete pearls—or an intertwined string following the double-helix metaphor—that are collected on a given chromosome. But Tjian reinforces the significance of the discovery that much less than half of the base sequences are actually coding for the creation of a protein.

He is searching for the messages contained in the larger (by a factor of three or four) "noncoding" portion of the human genome. Mapping is one thing: eventually with the aid of supercomputers and refined experimental and microscopy techniques to probe the DNA material, an army of researchers will have diagrammed a map that shows the generic, linear sequence of all of the nucleotide base pairs, which number about 3 billion in humans. For Tjian, however, that will only be like the gathering of a big pile of puzzle pieces. He is looking to the next stage, trying to put the puzzle together, but from this early point in the process it is difficult to say for certain even what the size and shape of the individual pieces look like. He knows the title of the assembled picture: "How the DNA governs all of the intricate complexities of life."

One early insight is proving important and echoes revelations uncovered by scientists studying complex and dynamical systems: each of the trillion cells in a human is not an autonomous entity that, once created by the DNA, operates like a machine. Life is a process, calling for infinitely many and infinitely subtle reactions and responses to the conditions that unfold. The formal way of clarifying this is to refer to the actual generic sequence of bases in an organism as its genotype, and the actual physical life form that genotype evolves into as the phenotype. The distinction assumes ever more significance when the effects of interacting dynamic phenomena on a system's evolution are considered. A popular slogan among biologists goes: evolution only provides the genotype; the phenotype has to pay the bills. Tjian and his colleagues strongly suspect that, on the cellular level—which is the level where molecular biologists quest and where DNA eventually produces its effects—the instructions are not merely laid down and then run out like a permanently and deterministically wound-up clock. Most of the noninstinctive—that is, other than biochemically exigent and predictable—functions performed within the cell must have a guiding intelligence, and that intelligence must be coded in the DNA. And these modern geneticists are most compelled by the very functions that begin the process, translating the DNA's program into action.

The Central Dogma of Biology

Not long after Crick and Watson made their celebrated discovery, they pursued their separate researches, and Crick was among those given the most credit for helping to unravel the code itself. In the process, it became clear that DNA was not really an actor at all, but rather a passive master copy of the life plan of an organism's cells. Crick was responsible for what came to be called the central dogma of biology—the sequence of steps involved in the flow of information from the DNA master plan through to the final manufacture of the proteins that power the life process (Figure 5.3).

Molecular biologists and biochemists have uncovered a number of fascinating and unexpected phenomena at each of these distinct steps. But the transcript made at the first step is understandably critical, because somehow the proper part of the enormous DNA master plan—the correct gene or sequence of genes—must be accessed, consulted, and translated for transmission to the next step. Thus the major questions of transcription—often referred to as gene expression—draw the attention of some of the world's leading geneticists, including Tjian and his colleagues at the symposium's gene regulation session, who explained how they probe the mRNA process experimentally in search of answers.

A cell has many jobs to do and seems to be programmed to do them. Moreover, the cell must react to its environment and thus is constantly sensing phenomena at its cell membrane with receptors designed for the task, and then transmitting a chemically coded signal to the nucleus. Processing this information, to continue the metaphor, requires a software program, and undoubtedly the program is located in the genes. It is the job of the transcription machinery to find the proper part of the genome where the needed information is located. Conceptually, two categories of signals may be received,

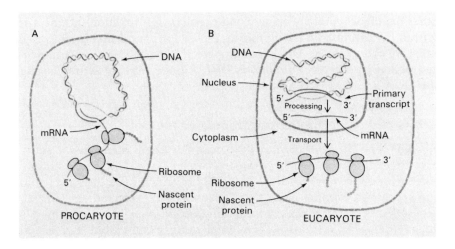

FIGURE 5.3 (Top) Pathway for the flow of genetic information referred to in 1956 by Francis Crick as the central dogma. The arrows indicate the directions proposed for the transfer of genetic information. The arrow encircling DNA signifies that DNA is the template for its self-replication. The arrow between DNA and RNA indicates that all cellular RNA molecules are made on ("transcribed off") DNA templates. Correspondingly, all proteins are determined by ("translated on") RNA templates. Most importantly, the last two arrows were presented as unidirectional; that is, RNA sequences are never determined by protein templates, nor was DNA then imagined ever to be made on RNA templates. (Reprinted with permission from Watson et al., 1987, p. 81. Copyright © 1987 by The Benjamin/Cummings Publishing Company, Inc.). (Bottom) Transcription and translation are closely coupled in procaryotes (A), whereas they are spatially and temporally separate in eucaryotes (B). In procaryotes, the primary transcript serves as mRNA and is used immediately as the template for protein synthesis. In eucaryotes, mRNA precursors are processed and spliced in the nucleus before being transported to the cytosol. [After J. Darnell, H. Lodish, and D. Baltimore. Molecular Cell Biology (Scientific American Books, 1986), p. 270.] (From p. 716 in BIOCHEMISTRY 3rd edition, by Lubert Stryer. Copyright © 1975, 1981, 1988 by Lubert Stryer. Reprinted by permission from W.H. Freeman and Company.)

though probably in the same form. One could be thought of as pre-programmed; for example, when a cell begins to die its natural death, it must be replaced, and a full new set of DNA must be created for the progeny cell. Such a DNA replication event is biologically predictable, and thus it could conceivably be anticipated within the program itself. But a different sort of signal is probably far the more numerous sort: a need to respond to something exigent, a reaction—to some extracellular event or to an intracellular regulatory need—that requires a response. With this latter sort of signal, the RNA-transcribing enzyme, RNA polymerase, is somehow able to search out the proper part of the DNA library where the needed information is stored, copy it down by transcription, and then deliver the transcript to the next step in the process, which will move it outside the nucleus to the ribosomes. These have been described as the production factory where the body's proteins are actually assembled using yet another variant of RNA, ribosomal RNA (rRNA). Again, the central dogma.

TRANSCRIPTION FACTORS

Tjian believes the key to unravelling the complexities of the code lies in understanding how the messenger RNA transcript is crafted. Since the chemical rules by which RNA polymerase operates are fairly well understood, he is looking for more subtle answers, related to how the protein finds the proper part or parts of the genome—that is, the gene or genes that need to be consulted at the moment. His research indicates that the answer most likely will involve at least several phenomena, but his target at the moment is a collection of proteins called transcription factors. Since timing is also a crucial component of the transcription process, geneticists are trying to understand how the rapid-fire creation of proteins is coordinated: not only where, but when. This is because the ultimate product, the long polypeptide chains that make up the body's proteins, are linear as they are built. This long string, when conceived as the product of a program, can be seen as the sequential order in which the proteins are called for and assembled, because they are strung together one after another by chemical bonding in a long chain in one direction only. The ribosomal cell factories pump out proteins at the rate of over 30 per second. An only slightly fanciful example: if the RNA polymerase is moving down the DNA chain, and at 1.34 seconds the code says UCA (serine), at 1.37 seconds it says ACG (threonine), and then at 1.40 it says GCA (alanine), there cannot be a delay in reading the UCA, or the proteins will not be laid down in the proper sequence, and the protein sequence and therefore the resultant polypeptide chain will be different, and the whole system will break down.

Thanks to the electron microscope, Tjian was able to provide moving pictures of the transcription process in action. Almost all the actors in this drama are proteins in one form or another; the primary substance responsible for making the transcript is a complex protein called RNA polymerase II. RNA polymerase II is a multisubunit enzyme composed of approximately 10 different polypeptides.

The first step is to clear the DNA strand of associated chromatin components so that the RNA polymerase can get at it. The DNA molecule in a eukaryote is wrapped up in a complex of proteins called histones, which have to be cleared off the DNA template. Then the two complementary strands are locally unwound, and the RNA polymerase starts to move along one strand to create the transcript. By reading nucleotide bases it is actually building a complementary strand of mRNA, by producing the base that chemistry calls for. The mRNA transcript is actually a copy—with the substitution of uracil for thymine, the RNA domain's one primary change—of the DNA's sense strand, which is merely standing by while the formerly coupled antisense strand, its chemical complement or counterpart, is being used as a template (Figure 5.4). The template does not produce

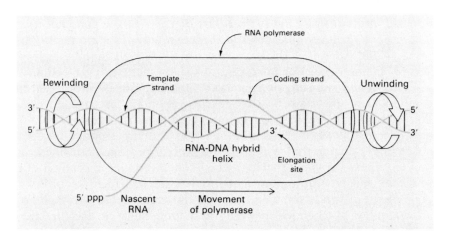

FIGURE 5.4 Model of a transcription bubble during elongation of the RNA transcript. Duplex DNA is unwound at the forward end of RNA polymerase and rewound at its rear end. The RNA–DNA hybrid helix rotates in synchrony. (From p. 710 in BIOCHEMISTRY 3rd edition, by Lubert Stryer. Copyright © 1975, 1981, 1988 by Lubert Stryer. Reprinted by permission from W.H. Freeman and Company.)

an exact copy but uses basic chemistry to create a complementary copy of the antisense strand, ergo an exact copy of the sense strand.

Tjian concentrates not on the chemical events themselves, but rather on how the RNA polymerase somehow knows where to go to begin and then where to end the transcript. The basic terrain he is exploring—the proteins called transcription factors—has signposts like promoter and enhancer regions, and introns and extrons, to guide this search for the "where" of transcription. The roles these regions play are largely unknown, and they offer a rich terra incognita for the molecular biologist and biochemist with a pioneering curiosity. As Tjian put it, "RNA polymerase is rather promiscuous" and not capable of discriminating discrete parts of the genome. It is the transcription factors that "seem to be designed to recognize very subtle differences in the DNA sequence of the template and can easily discriminate a real piece of information from junk. Their resolving power is very high," he said. Used as probes, they allow geneticists to home in on a piece of DNA as small as 6 to 8 nucleotides in length.

Drawing Lessons from Simpler Organisms

As Arnold Berk put it: "What are the punctuation marks?" Berk has done some important work on this question by probing a much simpler genome in a particular species of virus called adenovirus type 2, one of the 100 or so viruses that give humans colds. This diversity of known cold viruses allows Berk to deduce that "this is why you get a cold every year." Since a single exposure is sufficient to create a permanent immunity to its effect, it is comparatively safe to work with it in the laboratory. As compared to the 3 billion base pairs in the human genome, the adenovirus has only about 36,000. The logical inference is that—even though the virus does not have to perform differential calculus or ponder the philosophical and scientific implications of chaos theory—it has a more efficient genome. That is, there is a smaller proportion of junk or extra (that is, unidentified as to its clear function) DNA in the viral genome. "It grows well in the laboratory, it is convenient to work with," and its transcription behavior should be comparatively lucid.

The strategy of the virus, when it manages to get inside a host cell, is to exploit the cell's capacity to transcribe and translate DNA. The DNA of the virus says something like, "Make more of these." Instructions that say "make proteins" are obviously not hard to find, since they are the heart of any DNA code. But since it has been revealed that, in the human and most other genomes, so much other curious DNA is present, the virus cannot simply find a generic, "make more of these" instruction sequence. The virus must locate one or a series of very specific transcription factors that it can use to subvert the host cell's original genetic program for its own purpose, which is self-replication. If it can find the right place to go, it wins. For, as Berk says, "the cell cannot discriminate between the viral DNA and its own DNA, and it reads this viral DNA, which then instructs the cell to basically stop what it is doing and put all of its energy into producing thousands of copies of this virion."

Berk's experiment takes full advantage of the relative simplicity of the viral genome. The experimental goal is to elucidate in the DNA sequence a so-called promoter region—the "punctuation mark," he explained, "which instructs transcription factors and RNA polymerase where to begin transcribing the DNA." Because "some of these control regions that tell the polymerase where to initiate are very much simplified compared to the control regions you find for cellular genes," Berk has been able to home in on the promoter region for the E1B transcription unit. This process illustrates one of the basic genetic engineering protocols.

Because of the simplicity of the viral genome, he and his team began by narrowing the target area down to a point only 68 base pairs beyond a previously mapped gene region. To target even closer, they next constructed mutants by removing or altering small base regions on successive experimental runs. The process of trial and error eventually locates the precise sequence of bases that works, that is, initiates transcription in the host cell. The result is the clarification of a particular promoter region on the viral genome, more knowledge about the transcription factors that interact with this promoter region, and, hopes Berk, some transferrable inferences about the structure and function of more complicated promoter regions in animal cells.

Exploring the Details of Binding Protein to DNA

One of the first hurdles in performing laboratory experiments like the ones Berk and Tjian described is actually getting a sufficient quantity of transcription factor proteins to work with. "They are very elusive because they are produced in minute quantities in the cell, along with hundreds of thousands of different other proteins, and you have to be able to fish these particular proteins out and to study their properties," explained Tjian. However, since their structure includes a surface that recognizes a chemical image of a DNA sequence, experimenters can manufacture synthetic DNA sequences, by trial and error, that eventually match the profile of the protein they are trying to isolate and purify. These DNA affinity columns are then attached to a solid substrate and put into a chemical solution. When a solution with thousands of candidate proteins is washed past these tethered DNA strings that geneticists refer to as binding sites, the targeted transcription factor recognizes its inherent binding sequence and chemically hooks on to the probe. Once the transcription factor is in hand, it can be analyzed and duplicated, often by as much as a factor of 10^5 in a fairly reasonable amount of laboratory time.

Tjian illustrated another method of doing binding studies, that is, isolating the small region of the DNA sequence that actually contacts and binds the transcription factor. The first step is to tag the DNA gene region by radioactive labeling and then to send in the transcription factor to bind. Next, one tries to probe the connection with "attacking agents, small chemicals or an enzyme that cuts DNA." The bound protein literally protects a specific region of the DNA from chemical attack by these agents and thus allows the detailed mapping of the recognition site. "Gel electrophoresis patterns can actually tell, to the nucleotide, where the protein is interacting" (Figure 5.5).

After some years now of experience with binding studies, Tjian and other molecular biologists have begun to recognize certain *signatures*, structures that seem to indicate transcription factor binding domains. One of the most prominent are the so-called zinc fingers, actually a specific grouping of amino acids that contains a zinc molecule located between cysteine and histidine residues. Over and over again in binding studies, analysis of complex proteins showed Tjian this "recognizable signpost . . . a zinc finger," which, he and his colleagues surmised, "very likely binds DNA." Subsequent analysis showed that it was, in fact, usually embedded in the effective binding domain. In this very specialized area of research, Tjian called this discovery "an extremely powerful piece of information" that has led to the cataloging of a large family of so-called zinc-finger binding proteins. Another similar binding signature they call a helix-turn-helix, or a "homeodomain."

Using the Power of Genetics to Study Transcription

Biochemist Kevin Struhl described for symposium participants some of the various methods used in his laboratory's work with yeast, an organism whose relative simplicity and rapid reproducibility make it a good candidate for studies of the transcription process. "As it

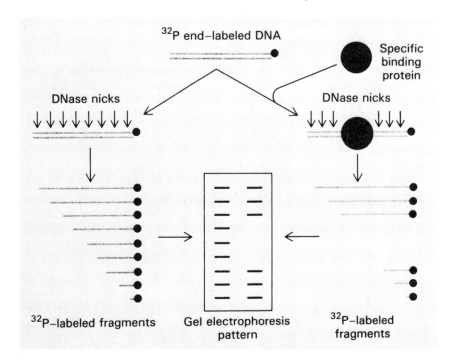

FIGURE 5.5 Footprinting technique. One end of a DNA chain is labeled with ^{32}P. This labeled DNA is then cut at a limited number of sites by DNase I. The same experiment is carried out in the presence of a protein that binds to specific sites on DNA. The bound protein protects a segment of DNA from the action of DNase I. Hence, certain fragments will be absent. The missing bands in the gel pattern identify the binding site on DNA. (From p. 705 in BIOCHEMISTRY 3rd edition, by Lubert Stryer. Copyright © 1975, 1981, 1988 by Lubert Stryer. Reprinted by permission from W.H. Freeman and Company.)

turns out," Struhl pointed out, "the basic rules of how transcription works are really fundamentally the same in yeast and in humans and all eukaryotic species."

One genetic approach researchers have used in yeast to try to identify some of the key proteins involved in transcription involves isolating mutants whose properties differ in some respect from those of the normal organism. "In yeast cells," he explained, "one can easily isolate mutants that do or do not grow under certain circumstances. . . . The aim is to try to understand the biology of some particular process, for example, a cell's response to starvation conditions," by identifying mutants that do not display a particular property and then checking experimentally to see if absence of a "normal" or unmutated gene accounts for absence of that property in the cell. "The basic idea," he continued, "is to first look at the function of an organism and to identify a variant that cannot perform it. Getting a function and a mutant is a first step in discovering which gene is actually involved in regulating the property being studied, and then in learning about the transcription factors that regulate expression of the gene."

Another method, gene replacement, was described by Struhl as a "very powerful technique to identify all the different parts of the gene and what it is doing." He explained: "To put it in simple terms, the process is analogous to going into a chromosome with scissors, cutting out a gene, and then replacing it with one that the researcher has created in a test tube." The result is "a real, intact cell . . . that can be analyzed to learn what the result of that gene is."

A third technique, developed in Struhl's laboratory and now widely used in the study of transcription, is one that he called reverse biochemistry. The researcher essentially carries out in a test tube what normally happens in a cell. Struhl pointed out that one "can actually take the DNA in a gene, use the appropriate enzymes to synthesize RNA and the protein [it encodes] in a test tube, . . . and then test to see exactly what the protein does." An advantage is that purification procedures required for work with unsynthesized proteins can be bypassed. In addition, a protein that is synthesized in a test tube can also be treated with radioactive label, which in turn enables many interesting related experiments.

A final technique mentioned by Struhl is used to figure out how much information there is in a particular genetic function, or, more specifically, how much DNA a specific DNA-binding protein actually recognizes," and what, really, is being recognized. DNA is synthesized so that the 23 base pairs are in a completely random sequence. Because DNA can be isolated only in minute quantities, "every single molecule [thus synthesized] is a different molecule in terms of its sequence," Struhl explained. A DNA-binding protein, GCN4, is put on a column and the completely random mixture of sequences is then passed through the column. Separating what is bound by GCN4 from what is not bound and then sequencing the result "gives a statistically valid description of what the protein is actually recognizing," Struhl said. In the case of GCN4, what the protein recognizes is the statistical equivalent of 8 1/2 base pairs worth of information. "The important point," Struhl summed up, "is that this random selection approach can be used for many other things besides simply DNA binding. . . . If [for example] you put this random segment of

DNA into the middle of your favorite gene . . . you can ask all kinds of questions about how much specificity, in terms of nucleic acids, is needed to carry out a particular function of interest."

Activation—Another Role of Transcription Factors

Tjian reminded the symposium audience that "transcription factors have to do more than just bind DNA. Once bound to the right part of the genome, they must program the RNA polymerase and the transcriptional accessory proteins to then begin RNA synthesis," and to do so, moreover, with exquisite temporal finesse. Experiments indicate that an altogether different part, as Tjian puts it, "the other half" of the transcription factor protein, does this, probably by direct protein-to-protein interaction, triggering regulation on or off, up or down. A powerful insight from these studies is that transcription factor proteins seem to consist of at least two modules, one for binding and one for activation. The modular concept has been borne out experimentally as to both structure and function. Molecular biologists have been able to create hybrid proteins, mixing the binding domain from one gene with the activation domain from another. Fortunately, signatures have also been detected that often indicate the presence and location of these activation domains. One such signature is a particularly high concentration within a certain protein of the amino acid glutamine, and another is a similar cluster of proline molecules. Though they do not know how this concentration may actually trigger the transcription process, geneticists have some confidence that the signature does encode for the activation domain of the transcription protein.

The accumulated value of these discoveries begins to suggest the figure from the ground. And though biology does not yet have a good model for how transcription factors do all of their work, a catalog of signatures is a vital foundation for such a model. Such a catalog "tells you two things," said Tjian, "first, that not all binding [and/or activation] domains use the same architecture, and second, that there is tremendous predictive value in having identified these signatures, for you can then say with some confidence whether a new gene that you may have just discovered is doing one of these things and by which motif."

These binding and activation domain studies also suggest another feature of transcription factors that Tjian referred to as topology. Even though these polypeptide chains may be hundreds of molecules in length and are created in a linear progression, the chain itself does not remain strung out. Rather it tends to coil up in complicated but characteristic ways, with certain weak chemical bonds forming the whole complex protein into a specific shape. When this complexity of shape is imposed onto the DNA template, experiment shows that a given protein—with a presumably specific regulatory function—may make contact with other proteins located at several different places hundreds or thousands of bases apart. As Tjian said, "Specific transcription factors don't just all line up in a cluster near the initiation site, but can be scattered all over." The somewhat astounding results of such topographic studies suggest that the molecules somehow communicate and exert "action at a distance," he said, adding that their effects are synergistic. That is, molecules at a proximal site are causing a certain effect, but when other distant molecules that are nonetheless part of the same or related transcription factor complex make contact, the activity at the proximal site is enhanced. Electron and scanning microscopy confirms these spatially complex interactions, the implications of which would seem to suggest a fertile area of inquiry into the overall process of transcription. As Tjian pointed out, "This gives you tremendous flexibility in generating a much larger combinatorial array of regulatory elements, all of which can feed into a single transcription unit. You can begin to appreciate the complexity and, also, the beauty of this transcription factor regulatory system."

Notwithstanding the complexity and continual reminders of what they do not know, geneticists have established some basic rules that seem to govern transcription activation. A line of exceptionally active and hardy human cells called He-La cells have proven very manipulable in vitro and indicate that in all transcription events, a "basal complex" must first be established. Many genes seem to present a so-called TATA (indicating those particular bases) box to initiate the binding process. The TATA box-binding protein alights first on the gene, and then another specific molecule comes along, and then another, in a characteristic sequence, until what Tjian called the "basic machinery," or basal complex, has been assembled. From this point onward in the transcription process, each gene likely has a specific and unique scenario for attracting specific proteins and transcription factors, but will already have constructed the generic, basal complex to interact chemically with them (Figure 5.6).

The Transcription Factor in Development

Thus far, most of the experiments described rely on the basic chemistry of transcription to indicate how it may work. Arnold Berk's mutant strategy, however, suggests that another mark of the effect of

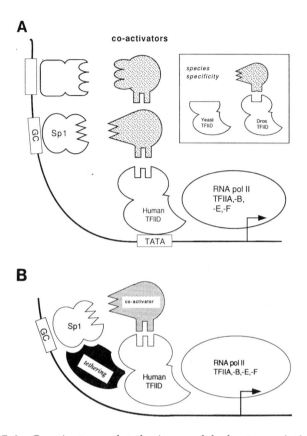

FIGURE 5.6 Coactivator and tethering models for transcriptional activation by Sp1. (A) A model for *trans*-activation through coactivators. This model proposes that specific coactivators (stippled) function as adaptors, each serving to connect different *trans*-activating domains into the general initiation complex, possibly to the TATA binding protein TFIID. These coactivators are not any of the basal initiation factors (TFIIA-TFIIF), which in this diagram are grouped together. The coactivator may be subunits of a larger complex that includes the TATA binding protein. Although not depicted by this model, the putative TFIID target may have multiple surfaces to accommodate coactivators from a number of *trans*-activators. (B) Tethering model for Sp1 activation of TATA-less templates. At TATA-less promoters, Sp1 requires a novel tethering activity (shown in black) distinct from the coactivators (stippled) to recruit the basal initiation factors. This model shows the tethering factor interacting with the TATA binding factor TFIID since its function replaces the TATA box and it copurifies with TFIID. However, the tethering activity could be interacting with other components of the basal transcription complex and bypassing the need for the TFIID protein. (Reprinted with permission from Pugh and Tjian, 1990, p. 1194. Copyright © 1990 by Cell Press.)

transcription factors is their success or failure as indicated by the genes they normally influence. These genes often code for characteristics that are clearly manifest in the species under examination, and if a mutant can be developed that at least survives long enough to be analyzed, important inferences about transcription can be developed. Several of the session's scientists described how genetics makes creative use of recombinant DNA technology to alter and study the expression of genes in the embryo and the developing organism.

Transcription factors are proteins that tend to influence the expression of genes in the mRNA stage. But since the point has been made that organisms—even at the cellular level—grow and evolve dynamically in a process that involves their environment, geneticists also look through the lens of developmental embryology to see if transcription factors may be involved in interactions other than simply those inside the cell nucleus.

Ruth Lehmann addressed the question of "how we get from a single cell to an organism" and revealed to the symposium some of the broad outlines of how genes may be turned on and off as the organism develops in the embryo and after birth. She and her colleagues study the fruit fly *Drosophila*, another favorite species for genetic experimenters. This species lends itself readily to experimentation because it has a manageable number of fairly distinct features, and is very hardy in surviving and manifesting some rather extreme mutations. "The complexity of the larval morphology is in striking contrast to the almost homogeneous appearance of the egg cell," Lehmann pointed out. After birth, the segments from anterior to posterior—head, thorax, and abdomen—are distinct.

Lehmann and her colleagues have asked how the undifferentiated fertilized egg cell uses its genetic information to create a segmented animal. Are distinct factors for the development of each body segment prelocalized in specific egg regions during oogenesis, or is a single factor, which is present at different concentrations throughout the egg, responsible for establishing body pattern?

By pricking the egg and withdrawing cytoplasm from various egg regions, Lehmann showed that two factors, one localized to the anterior and another concentrated at the posterior pole of the egg cell, establish body pattern in the head-thorax and abdominal regions, respectively. These factors are deposited in the egg cell by the mother during oogenesis. Genes that encode these localized signals were identified on the basis of their mutant phenotypes. For example, females that fail to deposit the anterior factor into the egg cell produce embryos that are unable to develop a head and thorax. Thus body pattern is established by distinct factors that become distributed in a concentration gradient from their site of initial localization, and these factors act at a distance.

One of these factors, "bicoid," which is required for head and thorax development, has been studied in the greatest detail in Christiane Nusslein-Volhard's laboratory at the Max-Planck-Institut in Tübingen, Germany. During oogenesis, bicoid RNA is synthesized by the mother and becomes tightly localized to the anterior pole of the egg cell. The protein product of bicoid RNA, however, is found in a concentration gradient that emanates from the anterior pole and spans through two-thirds of the embryo. Bicoid protein encodes a transcription factor that activates several embryonic target genes in a concentration-dependent manner; cells in the anterior of the embryo, which receive a high level of bicoid, express a different set of genes than do cells that are further posterior, which receive lower levels of bicoid.

The various studies on maternal genes in *Drosophila* show that no more than three maternal signals are required for specification of pattern along the longitudinal axis, and that one factor is necessary for the establishment of dorsoventral pattern. Thus a small number of signals are required to make pattern formation. The complexity of the system increases thereafter as each signal activates a separate pathway that involves many components. Although it is known that many of these components are transcription factors, it is unclear how these different factors work in concert to orchestrate the final pattern.

IS CANCER GENE-ENCODED GROWTH GONE AWRY?

Hanahan reminded the symposium scientists that "mammalian organisms are composed of a diverse set of interacting cells and organs." Like Lehmann, he is interested in connecting gene expression to how the organism develops and functions. "Often," he said, "the properties of individual cell systems are only discernible by studying disruptions in their functions, whether natural or induced." He is exploring abnormal development and disease, primarily cancer, using transgenic mice that pass on to newborn progeny an altered piece of DNA especially crafted by the genetic scientist in vitro. The next generation of animals can then be studied as the genetic expression of the altered DNA plays itself out over time during development— both in embryogenesis and as the animal matures. "Transgenic mice represent a new form of perturbation analysis," said Hanahan, "whereby the selective expression of novel or altered genes can be used to perturb complex systems in ways that are informative about their development, their functions, and their malfunctions."

The process once again utilizes the bipartite character of genes already discussed, namely that "genes are composed of two domains: one for gene regulatory information and one for protein coding information." Hanahan first prepares his manufactured gene, known as a hybrid. The gene regulatory domain he designs as it would occur in a normal mouse, but the protein coding domain he takes from an oncogene, so-called because it is known to induce cancer. Next, he removes fertilized eggs from a normal mouse, introduces his hybrid gene with a very fine capillary pipette, and then reimplants this injected embryo back into a foster mother that goes on to give birth to what is defined as a transgenic mouse, that is, one that is carrying an artificially created gene. When the transgenic mouse mates with a normal mouse, about half of the second-generation mice inherit a set of DNA that now includes this new gene, still recognized by its regulatory information as a normal gene but whose protein instructions code for cancer growth. As Hanahan put it, "Half of the progeny of this mating carry the hybrid oncogene. Every one of those dies of tumors. Their normal brothers and sisters live normal lives."

Beyond proving that oncogenes are heritable, and that only the protein coding portion is necessary to cause cancer in the offspring, Hanahan found some other suggestive patterns. First, although an affected mouse has this deadly oncogene throughout its DNA and thus in every cell of its body, only some of the cells develop tumors. Second, the tumors that do develop arise at unpredictable times during the course of the mouse's life. "From this we infer that there are other, rate-limiting events in tumors," and that simply possessing the gene does not predict whether and especially when a cell will develop into a tumor, Hanahan emphasized. All of the cells must be classified as abnormal, but they seem to undergo what he referred to as a sort of dynamic evolution as the organism ages. He has seen this phenomenon in several different environments. For example, even if all 10 mammary glands of a transgenic mouse express an oncogene, the offspring mice inevitably, and reproducibly, develop only one tumor, on average.

With an insulin gene promoter, he has observed the "cancer gene" expressed in all of the insulin-producing cells of the islets of the pancreas at 3 weeks, but only half of these islets begin abnormal proliferation 4 weeks later. At 9 weeks, another phenomenon is seen that Hanahan believes may be more than coincidental, that is, an enhanced ability to induce the growth of new blood vessels, called angiogenesis. Of the 400 islands of cells expressing the oncogene, one-half show abnormal cell proliferation, yet the percentage of full-blown tumors is only about 2 percent. Prior to solid tumor formation, a few percent of the abnormal islets demonstrate an ability to induce the proliferation of new blood vessels. Thus the genes that control angiogenesis become a strong suspect for one of the other rate-limiting factors that control cancer growth. Hanahan said these findings are "consistent with what we suspected from the studies of human cancers: while oncogenes clearly induce continuous cell proliferation, the abnormal proliferative nodules are more numerous than the tumors, and precede them in time. There is evidence that the induction of new blood vessel growth is perhaps a component in this latter process resulting in a malignant tumor." But angiogenesis is likely only one of at least several rate-limiting secondary events.

The dynamic evolution theory thus entails the premise that cells acquire differential aberrant capabilities as they mature. These differences could come from other DNA-encoded information in different genes altogether, but not until the hybrid oncogene is introduced to initiate the process does the system then evolve cancerously. Other suspected traits, like the ability to induce blood vessel growth, probably relate to dynamic phenomena of cells in normal development or in their actions. Some cancer cells seem able to ignore and trample over their neighbors, while others seem able to actively subvert their adjoining and nearby cells into aberrant behavior. Some cancer cells show a propensity to migrate more readily. Hanahan and his colleagues are looking at these phenomena and how they may be expressed. "Can we prove these genetically?" he asked, and went on to suggest that from such studies scientists hope to derive cancer therapy applications in humans. Although extremely suggestive, the implications are not yet clear. Cancer is a disease of uncontrolled cell growth. Many of the transcription factors have a role in regulating protein production. Many of these same transcription factors can act as oncogenes. Thus a key to unlocking the mysteries of cancer could be understanding in greater detail how the transcription factors actually influence the rate of protein production.

The oncogene studies provide yet another interesting and suggestive finding. Tjian reminds us that one of the important functions of genes is to provide information to respond to what amount to ecological crises for the cell. Most cells are equipped with receptors of one sort or another at their membrane. When some chemical or other physical stimulus arrives at a chosen receptor, a chain reaction begins in the cytoplasm of the cell in order to get the message into the nucleus, presumptively to consult the DNA master plan (by a process as mysterious as it is speculative) for a reaction. The routes through the cytoplasm are called transduction pathways, and it turns out that many of the transcription factors Tjian and others have been studying serve to mark these pathways. One particular protein is called AP1, which in other studies has been revealed as an oncogene. Said Tjian: "Nuclear transcription factors are also nuclear oncogenes. They have the potential when their activities or their functions are perverted to cause uncontrolled growth and neoplasia. The discovery that this family of regulatory proteins is actually an oncogene was greatly aided by the analysis of the yeast protein GCN4 that was largely the work of Kevin Struhl and Jerry Fink."

LIFE ITSELF

In less than four decades, standing on the platform erected by Crick and Watson and a number of others, the genetic sciences of molecular biology and biochemistry have developed the most important collection of ideas since Darwin and Wallace propounded the theory of evolution. To say that these ideas are revolutionary belabors the obvious: science is in the process of presenting society with a mirror that may tell us, quite simply, how to build and repair life itself, how to specify and alter any form of life within basic biological constraints. Recombinant DNA technology promises applications that pose fundamental bioethical questions. Taken together with the advances made in modeling the brain, these applications point to a future where natural organic life may one day become chemically indistinguishable from technology's latest model.

This future, however, is only a shimmering, controversial possibility. The geneticists at the Frontiers symposium were united in their humility before the hurdles they face. Discoveries continue to mount up. The halls of genetics, reported Eric Lander, are a most exciting place to be working, and that excitement has led to a unity of biological disciplines not evident a decade ago. But as Ruth Lehmann warned, cloning a gene is a far cry from figuring out how it works. It is that puzzle, seen especially through the lens of mRNA transcription, that she and her colleagues are working on. Whether it will remain a maze with ever finer mapping but no ultimate solution is for history to say, but the search is among the most exciting in modern science.

BIBLIOGRAPHY

Mitchell, Pamela J., and Robert Tjian. 1989. Transcriptional regulation in mammalian cells by sequence-specific DNA binding proteins. Science 245:371-378.

Pugh, Franklin B., and Robert Tjian. 1990. Mechanism of transcriptional activation by Sp1: Evidence for coactivators. Cell 61:1187-1197.

Stryer, Lubert. 1988. Biochemistry. Third edition. Freeman, New York.

Watson, James D., Nancy H. Hopkins, Jeffrey W. Roberts, Joan Argetsinger Steitz, and Alan M. Weiner. 1987. Molecular Biology of the Gene. Fourth edition. Volumes I and II. Benjamin/Cummings Publishing, Menlo Park, Calif.

RECOMMENDED READING

Beardsley, Tim. 1991. Smart genes. Scientific American 265:86-95.

Johnson, Peter F., and Stephen L. McKnight. 1989. Eukaryotic transcriptional regulatory proteins. Annual Review of Biochemistry 58:799-839.

Ptashne, Mark, and Andrew A.F. Gann. 1990. Activators and targets. Nature 346:329-331.

Sawadogo, Michele, and Andre Sentenac. 1990. RNA polymerase B (II) and general transcription factors. Annual Review of Biochemistry 59:711-754.

6

New Breakthroughs in Medical Diagnosis

Biology in general and medicine even more so provide a natural—if anthropocentric—setting to address the question, How do the fruits of scientific progress concern the individual? Medicine itself consists of two broad pursuits; first, the clinical—concerned with applying the art and science of medicine to healing people. A second focus of modern medicine is the research conducted to address more general scientific dilemmas whose resolution might contribute to remedies for disease, illness, and incapacity per se, in a larger context than that of individual patients. As practiced, for example, at the National Institutes of Health (NIH)—headquartered in Bethesda, Maryland, and regarded as the most comprehensive U.S. medical research organization—medicine necessarily involves a myriad of patients and subjects who undergo tests, protocols, and experimental treatments as medical researchers address the larger questions of etiology and treatment. The Frontiers symposium session on magnetic resonance imaging (MRI) provided a vivid illustration of how these two sometimes distinct aspects of medicine often merge.

In the last 15 years, MRI has moved from the embryonic research stages into the thick of modern clinical practice. Most of the specialists in the Frontiers session nearly every day see—if not dozens of patients—scores of MRI pictures to aid diagnosis. Keynote presenter William Bradley, Jr., director of the MR Center at Long Beach Memorial Medical Center, specializes in MR imaging of the brain and together with David Stark—according to session organizer Peter

Dervan of the California Institute of Technology—"wrote the book [Stark and Bradley, 1988]" on clinical MRI. Stark specializes in imaging the liver and abdomen, where contrast agents have been particularly useful for enhancing MR images. He teaches at Harvard University and also works at the Radiology Department of Massachusetts General Hospital. Both authors are practicing physicians whose primary charge is to produce superior pictures of patients who are referred by other doctors for MR imaging.

Graeme Bydder from the Hammersmith Hospital in London was another of the session's presenters and is, said Dervan, "one of the true pioneers in this area." Bydder explained how the general problem of motion in the area under consideration has led to some refined MRI techniques. John Crues III provides another example of the range of MRI specialists. Teaching on the faculty at the UCLA Medical School and also serving as director of MRI at Santa Barbara Cottage Hospital, Crues told about the increasing value of MRI in imaging the musculoskeletal system, hitherto thought to be the province of the x ray. If there was an exception to the generalization that clinical MRI involves its practitioners in a continual process of analyzing and refining the pictures made of individual patients, Robert Balaban, chief of the Laboratory of Cardiac Energetics at NIH, is it. His laboratory is building new prototype machines in an effort to elucidate some of the subatomic phenomena even most MRI practitioners take for granted.

MRI provides a new way to look inside the body. When compared to the earlier x-ray systems—even sophisticated current versions such as computerized tomography (CT) and positron emission tomography (PET)—MRI is noninvasive, and often dramatically superior in discerning details and subtleties. The enhanced safety and results derive from a different approach altogether. The x ray invades the body in a series of electromagnetic pulses that are targeted to collide with the electrons in the tissue under examination. MRI, however, rests on different interactions: first, since most MR imaging targets the nuclei of hydrogen atoms (^{1}H) found in water (H_2O), the interaction is with protons; second, the "disruption" caused by the radiation is far more benign, since the radio frequency (RF) pulses generated travel in wavelengths many orders of magnitude longer—by photons that have energy many orders of magnitude less powerful—than those in the x-ray range of the electromagnetic spectrum. Instead of targeting electrons and thus generating ions in the body, MRI manipulates the inherent magnetic properties of spinning protons, detecting information from how the protons react to the RF pulses and magnetic fields directed toward the target tissue.

Even though Wilhelm Roentgen's discovery of the x ray astounded the world and revolutionized medical diagnosis, its benefits were not entirely benign. The radiation used was in the higher ranges of the electromagnetic spectrum and posed hazards. As it collided with atoms in the body, sometimes glancing off, sometimes being absorbed and creating ions, the radiation produced a significant energy transaction. Also, notwithstanding the use of contrast agents, the process revealed no depth and therefore showed sometimes confusing overlap of images in the frame. By the 1970s, the computer's power to generate, analyze, and combine images had overcome this deficiency, and CT scanning produced dramatically improved and enhanced x-ray images in three dimensions. These views of the body enhanced the power of medical diagnosis significantly, and won a Nobel Prize in 1979 for one of the pioneers, computer engineer Godfrey Hounsfield.

PET scanning techniques also rely on interactions with particles, but the emanating source comes from within the body by way of radioactive molecules introduced there by nuclear physicians. These chemicals are designed—insofar as is possible—to go to the area under scrutiny, and once there to begin to decay and emit positrons. The positrons then collide—in some instances combine—with electrons nearby and produce gamma rays. This radiation emerges from the body to be read by detectors that can pinpoint the source of the rays.

Another recently developed imaging technique is Doppler ultrasound, in which the potentially dangerous high-frequency (short-wavelength) x rays and gamma rays are supplanted by the much lower-frequency (longer-wavelength) sound waves. The waves are directed toward the target and then analyzed as they emerge for the pattern of interference they then carry.

MRI practitioners approach the formation of an image with a fundamentally different assumption than targeting electrons or tissues for collision. The imager uses magnetization to affect a range of tissue and imparts magnetic energy to all of the ^1H protons in the target area. This contrasts with the x-ray process, in which increasing the (potentially harmful) amount of radiation is the only means for increasing the likelihood of impacts that lead to a successful image. Quantum physics also demonstrates that electrons vary far less in their ability to reveal density or spatialization than do water molecules in the body. Electrons are electrons, regardless of the atomic nuclei they are circling, and x-ray pictures often fail to discriminate very clearly between them. Water (and thus the presence of ^1H) in the target tissues varies much more significantly. Concentration, density, and even more subtle measures having to do with H_2O binding to macromolecules all provide possible ways that MRI can exploit variations in water in the body—even in very nearby and only slightly different tissue targets—to produce better pictures.

As Bradley described the intricacies of MRI, he clarified which particular tissues and conditions in the body yield the best MRI pictures. From a strictly clinical perspective, many tissues and processes are ideal candidates for MRI, though some are not. He and his colleagues touched on the reasons for some of these limitations but also suggested how the expanding horizons of MRI research are encompassing more and more clinical diagnostic situations. As more sophisticated methods of scanning, more discerning contrast agents, and altogether new techniques are refined, the potential clinical scope of MRI will expand.

In the world of modern clinical medicine, however, practitioners face a number of practical considerations. The machinery and staff for a comprehensive MRI center cost millions of dollars a year and patently must be supported by larger institutions or hospitals, many of which have not installed MRI centers. Clinical and cost considerations aside, some patients are not suitable candidates for MRI because they cannot tolerate the conditions of the procedure, physically or psychologically. Usually the patient must climb into a large tunnel, hold still for 10 or 15 minutes, and perhaps remain for two or three times that long for several series of images. The noise and isolation of the machines have also presented problems for patients. Thus MRI is but one of a number of approaches to imaging the body. X-ray procedures continue to be refined, sonogram imaging has many applications, angiography and arthrography give doctors a direct approach to the problem area, and there may always be certain situations where the MRI procedure is categorically impossible. And yet, if cost and other practical impediments can be overcome, the future value of MRI to diagnosis is almost without limit. As Bradley described it, MRI "really has become the most powerful diagnostic tool in clinical medicine since the discovery of the x ray."

Powerful, and also sophisticated. The science of radiology, following the discovery by Roentgen in 1895 of x rays, created a formal branch of medicine devoted exclusively to diagnosis. Medicine, of course, has for millennia involved its practitioners in the process of ascertaining what it was they were undertaking to treat. Said Hippocrates: "The capacity to diagnose is the most important part of the medical art." But not until the 20th century did that part become so distinct as to involve separate, specialized doctors and technicians whose relationship with the patient was conducted primarily through the imaging machinery. This development has continued as other imaging modalities were discovered and developed, and MRI technology continues this trend. As CT and PET scanning techniques have become more elaborate and computer-enhanced strategies more complex (both for obtaining and then for processing the images), specialists must develop more than the discernment to simply read the results. With MRI in particular, the choice of imaging strategies is fundamental to producing pictures with the hoped-for information. That choice, and the reading of the pictures, require on the part of the MRI practitioner a mastery of the physics and biology that underlie the process, a mastery that has not been so crucial with other imaging modalities.

Bradley has written, "Understanding the physical principles of MRI is often compared to understanding the operation of an internal combustion engine. One need not understand the intricate details of an engine to be able to drive a car. Similarly, one need not be conversant with the basic principles of nuclear magnetic resonance to understand certain simple aspects of an MR image" (Stark and Bradley, 1988, p. 108). However, he made clear in his presentation to the symposium that radiologists and other MRI practitioners must consider the various factors—both limiting and enhancing—inherent to the image they are after. These factors presume a fundamental knowledge of the principles of magnetic resonance, which he summarized for the scientists in his 1990 Frontiers keynote presentation, "Clinical Magnetic Resonance Imaging and Spectroscopy."

THE PHYSICS AND COMPUTATION BEHIND MRI

Like its predecessor the x ray, MRI grew out of studies in basic physics. Physics itself, however, was undergoing a dramatic growth into the new territory of quantum electrodynamics, leading to exploration of particle behaviors undreamed of even in Roentgen's time. The era just before the turn of the century would soon come to seem almost antiquated, after the views of Max Planck, Niels Bohr, Albert Einstein, and Max Born began to open up vast new horizons, though on a subatomic scale. For decades, much of the important work was done in America, as the European conflict spurred many scientists seeking both academic and personal freedoms to depart for American universities. Magnetic resonance pioneer Felix Bloch credited Otto Stern and I. Estermann with first elucidating the deflection of a molecular beam of hydrogen by an inhomogeneous magnetic field in 1934. I.I. Rabi in 1937 confirmed the magnetic moment of nucleons and then demonstrated that resonance could be induced using particle beams and further magnetic effects. The definitive conception came in 1946 when—in separate efforts by Bloch and Edward Purcell—magnetic resonance was demonstrated and explained in solids and liquids.

The field was then called nuclear magnetic resonance (NMR) spectroscopy and bore immediate, fruitful advances in chemistry, the study of magnetism, and eventually biology. Paul Lauterbur in 1972 solved the problem of transforming the spectra data to a three-dimensional image by manipulating the magnetic field itself; what was to become MRI he called zeugmatography. Medical applications have proliferated in the two decades since. Techniques and computer power have both become more sophisticated. State-of-the-art MRI machines now provide scientists with a wide arsenal of approaches to produce spectacular pictures of not only the details, but also the condition, of soft tissue, fluid flow, even bone. As Dervan mused, looking back on this history: "Whoever thought that that simple basic thread of research would ultimately lead to one of the most powerful clinical diagnostic tools now on planet earth? I think there are 2000 hospitals in the United States that are using this and probably 3000 worldwide."

How MRI Generates Information

"Magnetic resonance imaging is based on a harmless interaction between certain nuclei in the body (when placed in a strong magnetic field) and radio waves," began Bradley in his explanation of the physics involved. The process works only with nuclei having an odd number of protons or neutrons, and works better when the nuclei are in great abundance. The hydrogen atom provides the ideal MRI target for these two reasons: it has a single proton, and is the most abundant nucleus of any species in the body.

To summarize the process: the MRI operator designs and controls an *input* in the form of radio waves directed toward the ^1H protons in the target area. A further input involves the magnetic field environment applied to the area, where the area is usually limited to a thin slice of only several millimeters, and the information in such a voxel (volume element) is subsequently translated into a two-dimensional pixel (picture element) by the computer. The spin of all protons produces a characteristic magnetic moment, in this case one that is inherent to hydrogen. These spinning particles respond only to inputs adjusted according to the basic formula, called the Larmor equation: $\omega = \gamma \cdot B_0$. This equation defines the frequency of the RF pulse (ω) needed to produce resonance in the ^1H protons (which have a characteristic gyromagnetic ratio, γ) in a given magnetic field (B_0). When such a frequency is transmitted, those and only those nuclei absorb that predesigned energy input. "The Larmor equation provides the key to spatial localization of the NMR signal and subsequent image formation," emphasized Bradley.

The MRI practitioner influences the subsequent release of that energy by manipulating the magnetic field environment across a gradient so that only those protons at a very specific site are affected. As that energy is emitted, the protons return to their original energy state. The act of absorbing and emitting energy is repeated over and over again, very rapidly: the protons are said to be resonating, and they do so at the Larmor frequency. The resonance signal is picked up by a radio receiver tuned to that particular frequency. The information comes in the form of amplitude versus time. In order to produce an image, that information must be transformed into the domain of amplitude versus frequency. The Fast Fourier Transform (FFT) accomplishes this necessary conversion, as well as the task of distinguishing each proton's characteristic frequency from the composite signal.

Bradley explained the basic steps of the process in greater detail in order to clarify the factors he and other MRI practitioners consider when designing the inputs to target a particular area or tissue for imaging. The key to locating the precise area under examination is the nonuniformity of the primary magnetic field (which is created by the large, primary superconducting magnet). The MRI operator intentionally creates and designs this nonuniform field by superimposing over the main field additional magnetic fields produced by resistive electromagnets coaxial to the axis of the main field, continued Bradley, adding, "These additional electromagnets can provide linear spatial variation in the net magnetic field along the x-, y-, or z-axis, creating magnetic field gradients, g_x, g_y, and g_z."

From this spatial arrangement comes another dramatic improvement in diagnostic power over CT, because the CT image is limited to the plane of the gantry and produces clinical images in the axial and semicoronal planes only. With the complex magnetic fields that magnetic resonance imagers can establish, the practitioner can acquire images in any plane—axial, sagittal, coronal, or oblique. The equation $w = g_x x$, where $g_x = dB/dx$, explains how this is possible. The magnetic field strength through a combination of these various resistive magnets can be controlled at a precise point in three-dimensional space. Since protons in hydrogen will react to RF pulses that are a function of the field strength present (according to the Larmor equation), said Bradley, MRI experiments create "a one-to-one correspondence between frequency and position." Extremely precise localization is therefore possible.

To summarize: Within a given tomographic slice, magnetic field strength is established such that a ^1H proton at any minute point locatable on the x- and y-axes will resonate to RF pulses of a particular frequency, while ^1H protons at nearby and adjacent points will not absorb the RF pulses because their magnetic environment—and hence their Larmor frequency—is different. A series of slices are then each submitted to a similar MRI pulse sequence, and the z-axis is captured, allowing the computer to produce a fully three-dimensional image of any point in the body where hydrogen protons are free to resonate. Thus is the first MRI measurement obtained, which reflects hydrogen proton density. The ubiquity of water in and between almost all cells of the body explains why MRI can be directed virtually anywhere. However, just as all electrons are electrons, so too are all protons identical. And if their densities were the only information discernible, the MRI procedure would not have begun to revolutionize diagnostic imaging as it has.

At the heart of MRI are several other more subtle phenomena that derive from the quantum physics of resonance. As a generalization, MRI studies are conducted to explore how individual ^1H protons behave subsequent to their absorption of energy from the RF pulse. The protons are first distinguished by location, according to the field strength present. This feature explains only how a very precise location can be isolated in an image. The goal of MRI, however, is not to measure water or ^1H protons, but rather to produce revealing images of tissue and internal processes. Fortunately, these targets are almost invariably associated with water and can very often be distinguished in fine detail by how individual water molecules that are within the tissue behave. One way of describing this behavior is to ascertain the net effect a given ^1H proton's local chemical environment has on its otherwise predictable atomic activity—its return to equilibrium after absorbing the magnetization energy—compared to other protons nearby. Physicists classify as relaxation the primary behavior involved: over a measurable period of time, the ^1H atoms lose the energy that was absorbed from the photons of the RF pulse(s). In losing this energy, all of the protons that were energized by the RF pulse are involved in two basic activities, and single MR images result from a series of pulse sequences that are designed by the operator to focus on one or the other.

T_1 and T_2 Relaxation Times as Data

Bradley referred to the fact that each atomic species has its own characteristic magnetic moment. Where does this come from? The quantum quality of spin arises whenever a nucleus has an odd number of either nucleon. Even numbers of protons or neutrons cancel one another's magnetic potential, but when one of these particles is left unpaired, a potential magnetization arises (it possesses a quantum spin of 1/2) because it is spinning rapidly. In a large collection of such atoms, however, there is no net magnetization—species are only potentially magnetic—because the various small magnetic moments of millions of individual particles will statistically cancel one another out. When such a collection of atoms is put into the primary magnetic field of an MR imager, however, these particles are influenced and act something like atoms in a bar magnet, aligning with the magnetic vector of the field. In practice, however, not all of them align parallel to the field vector. "After a few seconds," Bradley explained, a slight majority of them align parallel to the field, the rest antiparallel. The difference is only about 1 proton for every 1 million magnetized, but "this excess gives the tissue [in which the ^{1}H protons are embedded] a bulk property known as its magnetization."

When exposed to a single RF pulse at the Larmor frequency, this magnetization, referred to as longitudinal, disappears. On the quantum scale, the population of protons has responded to the energy by equally distributing themselves parallel and antiparallel to the main field. Equilibrium, however, requires that an excess of 1 proton in a million point with the field, rather than an equal number with and against it. The excess protons that were flipped into the antiparallel direction are therefore in a higher energy state and tend to return. As they do so, they lose this energy, in the form of heat, to their environment. Physicists refer to this local region as the lattice, a holdover from earlier studies of relaxation in solids that often had a crystal lattice structure. Thus the phenomenon is called, variously, thermal, longitudinal, or spin-lattice relaxation. It is designated T_1 and is the time required for longitudinal relaxation to regain 63 percent of its equilibrium value. $1/T_1$ is the relaxation rate of a species.

A second phenomenon occurs as RF pulses are received by the ^{1}H target protons, however. The energy of this pulse, as any beam of energy will, provides its own magnetic moment, which can be seen to arrive at the main field from a perpendicular direction. "When the system is exposed to the oscillating magnetic component of a radio wave at the Larmor frequency, the magnetization will begin to precess about the axis of that second, smaller magnetic field at the Larmor frequency," explained Bradley. It will move from pointing along the z-axis rotating toward 90°, and continue down to the –z-axis at 180°, at which point the magnetization is said to be reversed. The rotation about the z-axis will continue, and the precession thus induces an oscillating magnetic signal. "An RF receiver coil placed such that its axis is perpendicular" to the main field, said Bradley, "will be sensitive to such fluctuating changes in magnetization, and a voltage will be induced, oscillating at the Larmor frequency. This signal is known as the free induction decay (FID)." Its exponential decay occurs with a time constant $T_2{}^*$.

In a perfectly homogeneous magnetic field, the magnetic effect of the RF pulse on transverse magnification would continue, once set into motion, for quite some time. This time, called T_2 relaxation time, can be calculated. In practice, however, the effect is observed to end much sooner than T_2. Why? "Differences in local magnetic susceptibility and imperfections in the magnet result in local hot and cold spots in the field, with resultant differences in magnetic frequency," explained Bradley. Each of the spinning protons producing its own little magnetic moment is exposed to a slightly different microscopic magnetic environment, which is itself changing rapidly. As an analogy, one wave will ripple smoothly outward on the surface of a lake, but a series of stones thrown in create multiple waves that eventually cancel out. So, too, do these protons having slightly different phase angles, and spinning at slightly different frequencies, cause the whole system to run down or dephase. The crucial factor in all of this atomic variation, however, is the tissue environment where these protons are being measured. In a given area, in the water associated with a group of similar cells—a tumor, for example—the relaxation rate of the protons will be similar enough to distinguish as a group from another nearby signal emitted by hydrogen protons in cells or tissue whose biological condition is different.

But, for a variety of reasons, T_2 is the preferred measurement. What good is T_2 if it only applies to homogeneous magnetic field conditions that cannot be attained? A major refinement in MRI was the development of the spin-echo technique, a way of purifying the signal so as to compensate for the irregularities of the field. It was established that when a given RF pulse is sent (calculated to rotate the magnetization by 90°), the decay of the signal is a function of $T_2{}^*$. If a second pulse of twice the duration is then sent (to rotate the magnetization an additional 180°), a so-called spin echo forms. This echo is seen to reach its maximum amplitude in exactly as much time as the period that elapsed between sending the first (90°) pulse and the second (180°) pulse. The time from the first pulse until this spin echo occurs is called the TE or echo delay time. The effect of this series of pulses is to bring the precession rates of the various randomly precessing protons back into phase, regardless of where they were at the time of the second RF pulse. Thus the spin-echo technique allows T_2 to be measured. It is also referred to as the transverse, or spin-spin relaxation, time.

T_1 and T_2 Time as a Diagnostic Lens

Thus T_1 is a measurement of how quickly a proton (in reality the average of millions of protons at a given site) that is resonating at the Larmor frequency recovers 63 percent of its equilibrium magnetization. T_2 measures how quickly this collection of protons at a given site loses 63 percent of its phase coherence, once that coherence is induced by a pulse sequence designed to coordinate the spinning motion of precession of the various particles. T_2 can never be longer than T_1, since once a particle has lost its resonant energy, phase coherence becomes irrelevant. In pure water in vitro, both T_1 and T_2 are about 2.7 seconds. As water becomes more complex with solutes, these times shorten, since the presence of other materials provides additional magnetic forces that speed up the energy drain and the phase interference. Generally speaking, the ^1H protons in water associated with tissues show T_1 rates about 5 times greater, and T_2 rates about 50 times greater, than the rates in pure water.

Exactly how these rates are affected by the environment of the protons is a question of manifold complexity, under continuing study, which dates back to the early days of NMR. The so-called BPP theory of relaxation unified work done by Nicholaas Bloembergen, Purcell, and R.V. Pound over four decades ago and provided a qualitative description of relaxation in solids and monomolecular solvents like water, ethanol, glycerol, and other oils (Fullerton, 1988, p. 42). But as NMR was being developed into MRI, the inadequacy of these concepts became manifest when applied to the extremely heterogeneous multicomponent solutions—consisting of very complex collections of proteins—that are found in human tissue. In particular, hydration-induced changes in the motion of water at or near the surface of macromolecules have a profound influence on relaxation times.

Bradley explained how the practitioner develops a pulse sequence—a series of RF pulses—based on a feeling for how the chemical environment in a given area will influence the local magnetic effects on the target protons. This general awareness, based on clinical experience, guides the design of a series of sequences intended to elicit proton density, T_1-, or T_2-weighted images (Figure 6.1). One variable is the period of time between 90° pulses, the repetition time (TR). When the spin-echo technique is used, another time parameter is the echo delay time (TE), which "is the time between the 90° pulse and

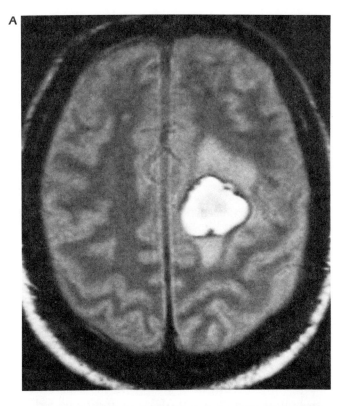

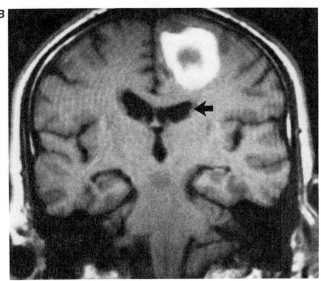

C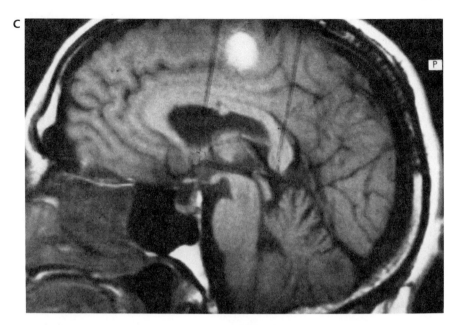

FIGURE 6.1 Multiplanar display capabilities for magnetic resonance imaging. (A) Proton-density-weighted (long-TR/short-TE) axial image through a chronic hematoma. This is the same plane visualized by x-ray computed tomography, which is parallel to the ground with the patient erect. (B) T_1-weighted (short-TR/short-TE), coronal plane through the hematoma demonstrates local mass effects, displacing the left frontal horn (arrow) inferiorly. (Coronal images are obtained in the plane of the face with the patient facing us.) (C) T_1-weighted image acquired in the sagittal plane, i.e., looking at the brain from the side in the midline with the nasal pharynx to the reader's left. By drawing a nearly horizontal line connecting the anterior and posterior commissures (the "bicommissural line"), and drawing perpendiculars to them, the central sulcus can be identified. Since the bright hematoma involves the precentral gyrus (the motor strip), we can predict on the basis of these images that motor deficits would be present as they were in the right leg. The combination of appearances on T_1- and T_2-weighted images allows us to determine that the central portion of the hematoma (which remains bright on both T_1- and proton-density-weighted images) contains extracellular methemoglobin, while on the proton-density-weighted image there is a dark rim surrounding the hematoma due to hemosiderin macrophages. This places the hematoma in the "chronic" class, which is at least 2 weeks old. (Courtesy of W. Bradley.)

the middle of the spin echo. Typically," Bradley continued, "tissues containing more water have longer T_1 relaxation times, while those containing lipid or paramagnetic species capable of a dipole–dipole interaction cause T_1 shortening." When the diagnostician examines the resultant images, "tissues with short T_1 relaxation times appear brighter, particularly on images acquired with short repetition times. Such short TR images tend to enhance the differences based on T_1 and are, therefore, known as T_1-weighted images," he explained. Conversely, because the longitudinal magnetization increases exponentially to a plateau, he added, "long TR images are thus said to be proton density weighted." When the spin-echo sequence is used, a similar variable is at work, and scans with a long TR and long TE are known as T_2-weighted images. Such T_2-weighted images appear brighter where T_2 is longer, as opposed to the T_1-weighted images, where shorter T_1 times produce a higher signal intensity and therefore a brighter image.

Looking More Closely at the Complexities of Relaxation

Robert Balaban and his team at the NIH are doing research into the question of how water molecules actually interact on or near the surface of the macromolecules that constitute so much of human tissue. They are examining extremely small magnetic interactions to explore how this mechanism might really work and to understand "why we get such good contrast with the different types of imaging sequences," explained Balaban. It was known that the relaxation time of a given proton depended on where it was located, its molecular neighborhood, so to speak. In pure water in the laboratory, T_1 and T_2 are identical. In tissue, they vary. The simple explanation is that the "freer the water, the longer the relaxation time." So-called bound water refers to water molecules tightly packed near protein structures classified as macromolecules. The "general dogma was that cross-relaxation or through-space interaction," said Balaban, is roughly "analogous to just getting two magnets close to each other. They don't have to touch, but of course they are going to interact through space as these dipoles get close enough."

Balaban's team wanted to see if this was really the primary cause of T_1 relaxation and also to come up with a description more chemically precise than just saying it was due to "cross-relaxation, or chemical exchange." They began with the basic methodology of saturation transfer but added a crucial variant. By setting the irradiation frequency only 5 to 10 kilohertz off the water frequency line, they were able to target the protons in the macromolecule without directly affecting the protons in free water (Eng et al., 1991). Thus any change in the water proton signal would be due to magnetization transfer between the macromolecule and water protons. The results clearly confirmed that the magnetization transfer with macromolecules was indeed the dominant source of T_1 relaxation for water in tissues. Moreover, since tissues vary as to their macromolecular makeup (and in turn also vary from the changes found in disease), this refinement yielded a predictable effect for each tissue type (Figure 6.2). What

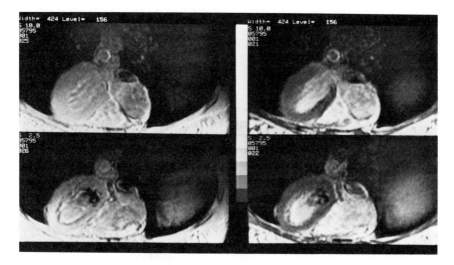

FIGURE 6.2 MRI image of the human heart collected at the National Institutes of Health on a normal volunteer. The images on the left are standard MRI images collected at different points in the cardiac cycle. The bottom of the image is the front of the chest in these axial images. The heart is in the center of the image with the large left ventricle on the left and the right ventricle below it to the right. The images on the right were collected under identical conditions, except that the macromolecules within the tissue were irradiated, revealing where water is in magnetic communication with these structures. A decrease in signal intensity occurs where the interaction is strongest; notice the interaction in the heart and skeletal muscle of the chest. The bright area in the center of the left and right ventricles is the blood within the heart chambers. This effect not only improves the quality of the MR image, but also promises to improve the diagnostic potential of MRI since this form of contrast is specifically linked to biochemical structures in the cell. (Reprinted with permission from Balaban et al., 1991. Copyright © 1991 by the Radiological Society of North America, Inc.)

was being measured was the rate constant for the interaction of a particular tissue's macromolecules with protons from adjacent water molecules. "So these studies seem to confirm that indeed, there is a relaxation pathway through the macromolecules in biological tissues, and," he continued, citing a large effect in skeletal muscles and very little effect in the kidney, "that it is a tissue-specific phenomenon." Thus by irradiating the sample with RF energy slightly off the expected Larmor frequency, Balaban and his colleagues devised a new method that produced heightened contrast where the macromolecular interactions were significant.

In clinical situations, it turned out that the presence of lipids in a given tissue was usually a dominant factor. Further studies suggested a very specific molecular reaction and isolated a specific OH group as the primary actor. What was being measured was the correlation time for the magnetic reaction in which the OH group was orienting the target water molecules. "It is not a chemical exchange," insisted Balaban, "but rather the action of an electric dipole" with a very long correlation time (10^{-9} seconds). "This model is much different from what we used to think about relaxation processes in biological tissues," Balaban summed up.

MAGNETIC RESONANCE IMAGING IN MEDICINE

Soft Tissues—Even Bone and Moving Joints— Are Prime MRI Targets

John Crues specializes in diagnosing musculoskeletal problems. When MRI began to reach clinical trials in the early 1980s, he reported, not many of his colleagues in orthopedics were optimistic that it would be of much use to their specialty because of an apparent major hurdle. Bones, explained Crues, "are made up of a matrix that is primarily calcium," an atom whose structure does not lend itself to imaging because it does not possess the mobile protons of hydrogen. Before long, however, surprisingly good pictures started to emerge (Figure 6.3) "because we had forgotten that the living elements of bone actually contain soft tissue that has a lot of hydrogen," said Crues. "When we started looking at the bones and the soft tissues of the musculoskeletal system with MR, we found that there really was a diagnostic revolution there," which he, too, compared to the vistas opened up by Roentgen's discovery of the x ray. Added Crues, "MRI has become the preferred imaging modality for the musculoskeletal system."

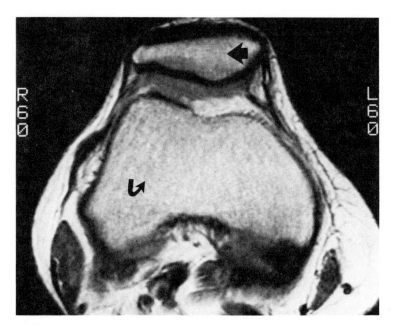

FIGURE 6.3 High-resolution MR image of the knee. This MR image shows the inside of a human knee painlessly and without the need for surgery. The patella (kneecap) can be seen anteriorly (fat arrow). The distal femur (thigh bone) is seen more posteriorly (curved arrow). The bright structures around the bones are fat. The dark structures are vessels, nerves, ligaments, tendons, and muscles. (Courtesy of J. Crues.)

Biomechanical joints such as the human knee and shoulder represent systems with a very precise design, produced by evolution for purposes that people living in modern times now routinely exceed. Modern athletes may not be that much more physically adept than their human counterparts who—a million years ago—roamed the savannas in search of food and shelter and in flight from enemies, but the persistent stress of modern activity puts their bodies and joints to a greater test. Medicine responds to the challenge of these greater demands by studying the joint systems, for example, in search of warning precursors, indicators of disease and breakdown. Until recently, such studies were limited to anatomical observations made during surgery and after death, when the system could be laid bare and its parts examined. The biomechanics were inferred from how these parts seemed to be connected and constructed. MRI and other modern diagnostics are now providing a much more scientific view of such systems. With contrast agents, the portable receivers Crues mentioned, and a version of MRI being refined that uses fluoroscopy, isotropic three-dimensional imaging can be done on a moving joint, functioning in real time under stresses designed by the diagnostician.

Twenty years ago, cartilage tears were often inferred from observing a characteristic presence of edema and interference with the range of motion. Mistakes were often made, however, given the complexity of the structure and the possible alternative kinds of damage that could be masked by the swelling. Although by the early 1980s the growing use of the arthroscope improved diagnosis significantly, a minor surgical procedure was still required just to confirm the diagnosis. But as MRI has been refined, it now reveals both anatomy and pathology, by "noninvasive diagnosis of joint and soft tissue injuries that was inconceivable 10 years ago. This has virtually eliminated diagnostic surgery in the musculoskeletal system," Crues continued, and now allows surgeons to devise a more definitive and sophisticated approach to specific pathologies. A compact and accessible area like the human knee allows physicians to use a variation on the large, body-size RF coils, which often produces very sharp pictures. Called a surface coil, this compact transmitter–receiver unit fits right over the knee itself, and by this close proximity to the target tissue improves the signal-to-noise ratio (SNR) and therefore image quality. Procedures have been developed so that diagnosis of meniscal and ligament tears is now performed easily in most cases.

Beyond enabling the fairly definitive evaluations of anatomy and pathology Bradley referred to, MRI gives Crues and other musculoskeletal MR specialists a tool with which to conduct proactive studies of the integrity of tissue in the knee. Torn or severed cartilage is a common injury, especially among athletes. Knee cartilage or menisci are small, curved tissue sacs that cushion the major ball-and-socket joint of the leg on either side of the knee joint. One of the underlying causes of meniscal tears is believed to be differences in the coefficient of friction between the top and bottom meniscal surfaces and the bones they slide against as the knee moves, said Crues. This pressure differential can cause shear stresses in the interior and tear collagen fibers, he continued, which the body responds to as it would to trauma or inflammation, leading to a "vicious cycle of . . . 'degenerative changes' and eventual mechanical breakdown." A physician who suspects the presence of such precursors to a rupture will design an MRI sequence to look for the longer T_2 that would be expected as the interior collagen fibrils uncoil and free up the water in and around their cells.

Crues believes that MRI goes beyond simple diagnosis, and he demonstrated how the views it provides have altered the fundamental medical approach to certain conditions. "We thought we understood these disease processes," he explained, until the subtleties and unprecedented detail of the MRI scans indicated complexities that were overlooked with previous diagnostic modalities. "One of the most common causes of knee symptoms in the population at large, especially in young women, is degenerative change in the articular cartilage of the patella [kneecap]," Crues continued. MR studies in the laboratory on this tissue revealed there are actually two layers of cartilage, usually with different water content. Studies over time of a given patient can reveal deterioration within these layers, and even a single MRI study may reveal more—since the cellular biochemistry revealed by MRI may indicate incipient damage—than would be discovered by a surgeon who cut in surgically for a look. "Further studies need to be performed to see if the signal changes [revealed with MRI] accompany early biochemical changes" that are known to be associated with diseased cartilage; if so, predicted Crues, "early detection may lead to changes in patient behavior, or to the development of drugs to protect the cartilage."

Tears of the rotator cuff in the shoulder provide another vivid example of how MRI is revolutionizing the treatment of joints. Such tears have many different possible causes, and in the past, cutting in for repair or removal did not allow surgeons "to distinguish among the etiologies," said Crues. "A major breakthrough in understanding this disease stemmed from the surgical findings of Charles Neer, who postulated that 95 percent of rotator cuff tears were due to chronic trauma," continued Crues, caused by a congenitally abnormally shaped acromion, the bone at the top of the shoulder. Normally this bone surface is flat, but "some people are born with a congenital abnormal shape," said Crues, a kind of bony hook that often abrades and produces chronic trauma on the rotator cuff muscle laid over it. Before MRI, this process was almost unavoidable in people with the condition. Now, said Crues, we "can adequately differentiate between those patients whom surgery may help, and those" whose shoulder joint degeneration stems from other causes. Now the area can be imaged, using short TE times and T_1-weighted strategies, and if the misshapen bone seems the likely cause, a quick arthroscopy to file it down will likely prevent altogether a tear that once—for lack of warning symptoms—was inevitable.

Unparalleled Views Inside the Skull

Perhaps in no subspecialty has MRI had such a profound impact as in neurology, where physicians understandably want as much information as possible short of cutting in for a direct examination. Until recently, the study of interior brain states was largely limited to the graphs of electrical activity discernible in electroencephalogram (EEG) studies.   With the development of CT scanning techniques over the last two decades, clinical neurology took a big step forward in the diagnosis and localization of problems.   Even more recently, PET studies (such as those that send a solution of glucose molecules labeled with fluorine-18 atoms to the brain) have provided real-time pictures of local activity as various regions of the brain metabolize other molecules for the energy to transmit their electrical signals. But, in many specifics, MRI has gone beyond even that level of detail. "With few exceptions, the sensitivity of MRI is greater than that of CT," said Bradley.

The brain is a very densely packed and convoluted system with some 10 billion constituent parts, its neurons. The vast terra incognita at the frontiers of brain research involves the intricate network of interconnections between these nerve cells, and how this network actually functions. But for clinical neurology, MRI can give surgeons and diagnosticians like Bradley extremely sophisticated information about the brain's anatomy, fluid flow, and the presence of hemorrhage, edema, and neoplasm. "Because of its sensitivity to small amounts of water, MRI is more sensitive to edematous lesions" than is CT, said Bradley. It can locate a lesion more precisely, and it "better demonstrates subtle mass effect, atrophy, and the presence of hemorrhage . . . and is currently the most sensitive technique available for detecting early brain tumors, strokes, or the demyelinated plaques in multiple sclerosis."

Although the brain is undoubtedly the most complex structure known to man, its component parts are not mysterious. MR images of the brain can distinguish clearly between the white matter, gray matter, the lipid-rich fat surrounding the scalp, and the cerebrospinal fluid (CSF). When tumors develop, the water they contain (and thus the protons emitting an MRI signal) is distinguishable from that in adjacent tissues. The CSF is largely water, and therefore an excellent target. One brain disease, hydrocephalus, can result from overproduction of CSF, but much more commonly from blockage (usually by a tumor) of the ventricles, the pathways allowing "old" CSF to flow out of the brain. "Any edematous process, any water-forming process in the brain or anywhere else in the body," explained Bradley, "increases the T_2 relaxation time, slows the rate of decay, and causes these processes to be highlighted on T_2-weighted images." This feature of MRI provides a nearly custom-designed camera for studies, not merely of the brain's gross anatomical structure, but of the condition of brain tissue as well.

Because MRI allows physicians to look directly at biochemical effects and cell processes, many strategies have been developed, including new approaches to internal bleeding. Explained Bradley: "Hemorrhage has many different forms. We can be fairly specific about the age of hemorrhage based on its MR appearance. Acute hemorrhage is primarily deoxyhemoglobin, which has a short T_2 and thus causes loss of signal. Typically, some days later, methemoglobin will be formed" and emit a different but characteristic signal that has "high intensity on any sequence and is therefore easily detectable."

The short T_1 time associated with methemoglobin can be attributed to the fact that it contains iron, which reacts directly with the surrounding water protons in a dipole–dipole interaction. Deoxyhemoglobin can be clearly distinguished from methemoglobin, said Bradley, because its associated "water molecules are unable to approach the heme iron within 3 angstroms for a dipole–dipole interaction," the mechanism whereby T_1 is shortened.

Paramagnetic substances act as contrast agents to enhance MR images. One such solid is ferritin, a "first cousin," explained Bradley, to "the hemosiderin that is a breakdown product of hemorrhage. Seeing this, we know that the only way it could have gotten there was from a prior bleed." Both of these show up as a conspicuous blackness due to short T_2 relaxation times. "Early in the subacute phase of hemorrhage," continued Bradley, methemoglobin within the red cells is magnetically very susceptible when compared to the plasma just outside these cells, leading to a short T_2. Later in the subacute stage, lysis of the red cells leads to a long T_2 and high signal. Thus variation in the T_1 and T_2 relaxation times indicates the sequential stages that may develop—and between which it is important to distinguish—in intracranial hemorrhage.

Capturing Fluid Motion with MRI

"Few aspects of MR are as potentially confusing as the effect of motion on the MR image," wrote Bradley (Stark and Bradley, 1988, p. 108). But the detection of motion "has no correlate in CT" scanning, and so pictures of flow represent a fundamental advance in diagnostic imaging. An invasive diagnostic x-ray technique called angiography has been developed to study blood vessels, but early indicators suggest MRI could largely supplant it, if techniques and cost considerations can be harnessed. "The routine angiogram requires sticking a needle about the size of a Cross pen into your groin, your femoral artery, where you feel the pulse, and then sending a catheter about a meter long up into the neck and injecting contrast into your carotid artery. It is not a pleasant procedure," commented Bradley, who has contributed to the development of "a painless procedure, MR angiography, that gives much the same information" in a matter of a few minutes, compared to the hour patients normally endure with the invasive and painful angiogram.

MRI studies can be directed not only toward the flow of blood, but also to the CSF in the central nervous system. Because of the abundance of protons in hydrogen and of water in the human body, MRI can detect and discern the movement of water anywhere in the body and is sensitive even at the smaller scales of molecules. MRI techniques are especially sensitive to the rate of flow. Blood traveling upwards from the heart through the aorta moves at nearly a meter a second. As it continues into the vascular network and moves into continually branching arteries, velocity is decreased proportional to the cumulative cross-sectional area. In major arteries in the brain, for example, blood is moving at about 10 centimeters per second. It continues to diffuse through the ever finer network until, reaching the capillaries, it is moving at about a millimeter per second. Generally, the more rapid the flow, the darker the MRI image: thus blood moving more slowly through the veins is lighter, and that pulsing through the arteries is darker on the image. Another convention used when imaging flow is to code for direction, which allows easy discrimination between inflowing and returning blood and CSF.

Before MRI practitioners could produce images of flow, techniques had to be developed to overcome a number of obstacles. Bradley pointed out that several "independent factors can result in decreased signal intensity of flowing blood: high velocity, turbulence, and odd-echo dephasing." Sometimes known as flow void, this decrease in signal can per se be revelatory, indicating under certain conditions the presence of aneurysms and arteriovenous malformations. To explain the mechanism behind the flow void, Bradley again described the spin-echo technique. Because the MRI equipment has been refined such that a thin slice on the order of 5 millimeters (or thinner) can be isolated for imaging, the speed of protons moving through the slice in a normal direction is problematic. "To give off a spin-echo signal," wrote Bradley, "a group of protons must be exposed to both a 90- and 180-degree RF pulse" (Stark and Bradley, 1988, p. 108). If the slice at the appropriate resonant magnetic field is thin enough, and/or the movement of the fluid fast enough, the protons may not be within the slice long enough to receive the two pulses and to produce the spin-echo signal.

The situation, however, still contains much inherent information for the MRI practitioner, explained Bradley. Flow that runs through the slice in a normal direction, rather than along it, will experience the flow void for the following reason. Certain protons will be entering the slice from the top and others leaving the slice from the bottom during the time of the MRI exposure. These will all receive and react to portions of the spin-echo sequence depending on their position, thus revealing information about direction and speed of flow when multislice analysis on the computer is accomplished. Another potential image problem concerns blood flowing essentially parallel to, rather than through, the slice. Since these protons will experience a changing magnetic gradient between pulses, they will not experience the coordinated effect of the pulses. One of the techniques developed to clarify these pictures involves elimination of the cyclical pumping effects of the heart. This is accomplished by gating, a technique that compensates for the cycle by timing its image exposures such that the same moment in the pump cycle is captured for each successive heartbeat. So, by correlations of phase, and subtracting images one from another, real-time pictures can be filmed, and fluid flow can be coded by the computer to reveal both its rate and direction. Flow patterns within the blood vessel can also lead to the void. Laminar blood flow results when blood near the center of a vessel moves faster than blood near the walls, causing a so-called slippage of layers. Turbulence, a form of chaotic and hence unpredictable motion, may also develop when blood flows rapidly through expanding vessels and changes in direction.

Just as some situations produce a tendency to decrease the MRI signal, "three independent factors also result in increased signal intensity," wrote Bradley: "flow-related enhancement, even-echo rephasing, and diastolic pseudogating" (Stark and Bradley, 1988, p. 108). Flow-related enhancement occurs when the spinning protons first enter a slice and have full magnetization. If an MRI sequence employs a short TR (i.e., is T_1-weighted), the protons in the flowing fluid have higher signal than the partially saturated (demagnetized) surrounding tissues. If a gradient echo technique is employed and only single slices are examined, "every slice is an entry slice," explained Bradley, referring back to the description of fluid flowing through, or normal to, the area under examination. This technique and the effect of "flow-related enhancement have recently been combined with a maximum-intensity, ray-tracing algorithm to provide the three-dimensional images of blood vessels" used in the MRI angiogram referred to above.

Contrast Agents

As David Stark observed, "Roentgen himself introduced the first contrast agents," and thus the principle of using contrast to enhance diagnostic images is as old as the x ray itself. "The general term" for the class of materials used, said Stark in his presentation to the symposium's scientists, is *"diagnostic contrast media."* They are pharmaceuticals that alter tissue characteristics [so as] to increase the information on diagnostic images. Each of the diagnostic imaging modalities—nuclear medicine, ultrasound, x-ray-based techniques—has its own family of dedicated contrast media."

The contrast agents used in MRI in certain situations can improve accuracy, reduce the number of possible diagnostic procedures, or shorten the course (and therefore the cost) of an MRI sequence. One significant distinction between MRI and CT contrast media is that with MRI the chemical agent is not itself being imaged. Rather, it produces an enhancing effect on one of the other MRI parameters already discussed, that is, relaxation. Stark pointed out that the key is to give a drug that goes to normal, but not abnormal, tissue and increases the signal difference between the two, thus giving a primary diagnostic test, which is to identify the abnormality as a result of the distribution of the drug. The fundamental point, he emphasized, is that these drugs can help us only if they target—or physically localize in—either the normal or the abnormal tissue but not both.

The only paramagnetic agent fully approved for clinical use by the Food and Drug Administration at this writing is gadolinium-diethylenetriaminepentaacetic acid (Gd-DTPA), although oral iron compounds and molecular oxygen also are being used in clinical studies. "In fact, some 30 percent of the patients who undergo NMR exams worldwide receive the drug," Stark said. Calling it, therefore, "a big ticket item," Stark reported that a number of corporations would like to break into this market, which "leads to some interesting issues in academia with sponsored research."

Stark explained that paramagnetic agents like Gd-DTPA serve to increase the parallel magnification of materials placed in the main field. "Paramagnetism is characterized by independent action of individual atomic or molecular magnetic moments due to unpaired electron spins" and "can occur in individual atoms or ions, as well as in collections of atoms or ions in solids such as ferritin and hemosiderin" and other substances in the body. Gadolinium-DTPA's potency is a measure of how much the proton relaxation rate is speeded up by a certain concentration of the drug. Gadolinium, which has a magnetic effect orders of magnitude larger than the effect of hydrogen nuclei on one another, provides a "built-in amplification step," said Stark, and can significantly shorten T_1 and T_2 relaxation times, T_1 more than T_2.

Since "tissue T_1 relaxation is inherently slow compared with T_2 relaxation, the predominant effect of paramagnetic contrast agents is on T_1," explained Stark. Gadolinium-DTPA, indeed all paramagnetic contrast agents, as they shorten T_1 serve to increase the signal intensity in the tissues that fall within the magnetic field gradient. When TE is short, T_1-weighted spin-echo and inversion recovery techniques tend to "show the greatest enhancement," he said. And with an increased signal-to-noise ratio in the image, scan times can be reduced.

An important characteristic of these agents is their biodistribution, how they tend to accumulate in the target tissue. Gadolinium-DTPA is administered intravenously and thus moves through the vascular system. When it reaches a particular tissue, differentiation may be observed because of a characteristic perfusion, relative to adjacent tissues, and likewise by diffusion through the capillaries into the extracellular spaces in the local region. Thanks to the blood-brain barrier, Gd-DTPA will only arrive at central nervous system tissue that is abnormal. Conversely, a properly functioning renal system tends to collect low-weight molecular compounds such as Gd-DTPA (590 daltons).

Stark also described another specific subset of drugs that has been designed for MRI enhancement. "These drugs work wonders in the brain and kidney," explained Stark, because—as small molecules—they are generally excluded by the blood-brain barrier, "and pathology selectively accumulates them, and the kidney handily excretes them." Thus any problem or pathology in the blood-brain barrier will be tagged by the use of Gd-DTPA, and any pathology in the kidney likewise will accumulate it. This picture of how the agent arrives at the target area also describes enhanced CT imaging, but the image contrast produced is usually superior with MRI.

Most uses of Gd-DTPA involve comparing the same MRI imaging sequence with and without the contrast agent. Contrast imaging is also indicated when a first MRI result indicates the presence of something not clearly definable, for which greater contrast is desired in a subsequent sequence.

With all contrast media, clinicians are concerned about toxicity. One problem inherent to some of these contrast media, said Stark, is that once introduced, they diffuse. "They pickle the entire body, just like most drugs that you eat or are injected with—they go everywhere. The whole body gets exposed to them even though the therapeutic or diagnostic effect is [confined to] one organ and represents a small fraction of the drug" that was administered. In all organs, the goal is to design an agent that will react differentially between normal tissue and the tumor. Thus researchers looking at possible new MRI contrast agents are searching for a "silver bullet," a substance that will go directly and exclusively to the problem organ by utilizing monoclonal antibodies.

Often referred to as the body's housekeeper, the liver, he pointed out, performs a similar kind of chemical discernment on many of the body's products. The liver hepatocytes take up some materials that a liver tumor cannot, and thus some liver cancers are easily detectable. Manganese-DPTP has been developed in a form that will go selectively to the liver and heighten the signal intensity in a T_1-weighted image, since it will not be present in the tumor. Researchers hope that a very precise and affordable test for liver imaging can be developed that will permit mass screening of the population for cancer. Also, such a process could provide insights that might lead to the development of additional contrast agents that would target selected areas of the body.

Spectroscopy and Diffusion Weighting

For decades, chemists have been using NMR principles to study atomic phenomena. With magnetic resonance spectroscopy (MRS), practitioners are beginning to explore the body through analysis of its chemistry in vivo. The process is similar to MRI, since the characteristic data come in the form of a unique resonant frequency, a response to input energy in the form of RF pulses rather than the laser light used in standard chemical spectroscopy. As with standard spectroscopy, however, each element in the periodic table as well as any variant with an alteration in the atomic composition of a molecule will produce a unique fingerprint. Therefore, even a single displaced electron will alter a target atom sufficiently to produce a slightly different resonant frequency, referred to as the chemical shift. Such a change in the atomic configuration will also alter the local magnetic neighborhood, so to speak, though the change in magnetism is very small compared to the power of the gradient field used in standard MRI. Since MRS is designed to detect these very small magnetic effects, the process requires an external magnetic field with much greater homogeneity (to within better than 0.1 ppm over the region of interest). Chemical shift information is often used as an additional refinement to further clarify pictures produced through standard MRI.

In the body, ascertaining the presence of certain substances at a particular site can often provide important indicators of disease or abnormality. Inositol aspartate seems to be present in all neurons and can serve as a marker for mapping the nervous system using MRS. Choline and creatine are associated with much new growth and with tumor production, and can therefore provide useful targets in spectroscopic studies. N-acetyl aspartate acid (NAA) is another useful marker for neurons and demonstrates an intriguing use for MRS in tracking the course of disease. "It is possible to create maps of [the brain using] NAA," said Graeme Bydder, by first generating "a normal one as a baseline" and then a subsequent MRS map "to observe neuronal fallout, for example, in Pick's disease. This allows us to detect the absence of neurons even where the MR imaging is normal." Since NAA serves as a marker for neurons, MRS studies of newborns over time can map the growth of the brain, and possibly detect abnormalities that would show up in no other tests.

Bydder's latest research interest involves imaging the movement in vivo of anisotropically weighted water, what he described as "diffusion-weighted imaging," a very novel approach that Bydder nonetheless relates to the field: "Ideas in magnetic resonance have come from a lot of different sources, and this is another one that has actually been borrowed from spectroscopy." Bydder credited Michael Mosley from the University of California, San Francisco, with the important work necessary to turn the idea into a clinical application. The motion of the water molecules due to diffusion causes signal loss that allows MRI to differentiate particular tissues. The anisotropic quality means that water protons diffuse at different rates from different directions. When this diffusion factor is added to the other MRI parameters, certain biochemical phenomena in the body may be revealed that would not show up in a straightforward search for differences in proton density or in T_1 or T_2 times. Tissues can in this way be distinguished by their "diffusibility," a further marker to discriminate fine detail.

Bydder called the use of this new technique in brain studies "an interesting area. It brings together the molecular motion of water, the properties of myelinated nerve fibers, the gross anatomy of the brain in terms of being able to distinguish the directions of white-matter fibers, and the diffusion of water" as influenced by the presence of disease. It is MRI that unites all of these factors, he concluded, which could prove to be a revolutionary diagnostic advance. Bydder cited improved diagnosis of multiple sclerosis as another application.

Previously designed MRI studies of the disease revealed a certain level of detail that he and his colleagues had agreed was "the gold standard. When we apply this [new form of] imaging, we can see a lot of extra abnormality" and detail that was not discernible before. "We have got to conclude either that we have a whole lot of false positives here," said Bydder, or that the new technique is more sensitive than the standard one. The realization that the anisotropic diffusion of water could yield this deeper look is so recent that it is now under intensive investigation, and new revelations may be expected. What is certain, however, is that MRI researchers will continue the search for techniques that will yield ever finer detail, which in these experiments has already reached the level of microns (one millionth of a meter).

PROSPECTS

Over 2000 systems are up and running in America, another 1000 elsewhere. Their average price was closer to $2 million than $1 million, and since much of that cost is associated with the magnet system, an MRI machine is going to be costly to procure and maintain until economies in the manufacture of superconducting systems are achieved. In evaluating this investment, hospitals will inevitably think in terms of "patient throughput," or the number of procedures per unit of time, at what billable return of costs. Another cost factor is the system's complexity. Radiologists and technicians, simply put, need to be better trained and qualified than with other imaging systems. But while medicine is a business, it is also much more. The growth and role of MRI will undoubtedly be determined more by physicians than by accountants and insurance companies. To the extent that MRI becomes essential to the provision of life-enhancing quality of care and to saving human life, to this extent will it assume its proper place in the medical armamentarium.

There are, however, some very interesting and promising possibilities, many of which were mentioned by the session's presenters.

• *Fast scanning techniques.* Because the utility of MRI in part depends on its ability to pay its way, faster scan times could lead to more images at a lower cost. These shorter imaging times will, in turn, increase the number of patients who can remain still and hold their breath for the duration of the process, and will simplify many of the longer gating procedures now required. Some in vivo work contemplates subsecond scan times. This is an area where funding for basic science, R&D, and theoretical studies could have a significant payback.

• *MR angiography.* As described, current traditional angiography techniques are extremely invasive. As MRI flow-study techniques are perfected, the former system may become antiquated.

• *NMR spectroscopy.* A number of biomedically interesting nuclei are being investigated to delve even more deeply into specific biochemical and intracellular processes.

• *MRI guided/assisted interventional diagnosis and therapy.* At present, CT and ultrasound are occasionally used in operating theaters to better control biopsy, drainage, and retrieval procedures. Significant physical obstacles would have to be overcome, but a system based on MRI could provide enhanced guidance and analysis.

• *MRI fluoroscopy.* Patients move certain joint systems—the jaw, the knee, the hip, the spine—through a graded series of positions, allowing MR imaging to produce studies of movement hitherto impossible. An analogous procedure is under investigation for cardiac studies.

• *New contrast agents.* A host of oral and intravenous contrast agents are being developed and tested.

• *Clinical efficacy.* MRI in many diagnostic situations has dramatically improved medicine's reach, penetrating even to the interior of the body's cells. Clinical horizons are expanding dramatically, and enhanced diagnosis of many areas and processes seems inevitable. Cost and other practical considerations are clearly an important element in this future. But the gold standard is as clearly in the process of being redefined.

BIBLIOGRAPHY

Balaban, R.S., S. Chesnick, K. Hedges, F. Samaha, and F.W. Heineman. 1991. Magnetization transfer contrast in and MR imaging of the heart. Radiology 180:367-380.

Eng, J., T.L. Cecker, and R.S. Balaban. 1991. Quantitative 1-H magnetization transfer imaging in vivo. Magnetic Resonance in Medicine 17:304-314.

Fullerton, Gary D. 1988. Physiologic basis of magnetic relaxation. Pp. 36-55 in Magnetic Resonance Imaging. David D. Stark and William G. Bradley, Jr. (eds.). Mosby, St. Louis.

Stark, David D., and William G. Bradley, Jr. (eds.). 1988. Magnetic Resonance Imaging. Mosby, St. Louis.

RECOMMENDED READING

Bradley, W.G., and G.M. Bydder. 1990. MRI Atlas of the Brain. Martin Dunity Publishers, London.

Crues, J.V. (ed.). 1991. MRI of the Musculoskeletal System. The Raven MRI

Teaching File. R.B. Lufkin., W.G. Bradley, Jr., and M. Brant-Zawadski, (eds.). Raven Press, New York.

Lehner, K.B., H.P. Rechl, J.K. Gmeinwieser, A.F. Heuck, H.P. Lukas, and H.P. Kohl. 1989. Structure, function, and degeneration of bovine hyalin cartilage: assessment with MR imaging in vitro. Radiology 170(2):495-499.

Mink, J.H., M.A. Reicher, and J.V. Crues. 1987. Magnetic Resonance Imaging of the Knee. Raven Press, New York.

Stark, D.D., and W.G. Bradley (eds.). 1992. Magnetic Resonance Imaging. Second edition. Volumes One and Two. Mosby, St. Louis.

Stoller, D.W. (ed.). 1989. Magnetic Resonance Imaging in Orthopedics and Rheumatology. Lippincott, Philadelphia.

7

Beyond Theory and Experiment: Seeing the World Through Scientific Computation

Certain sciences—particle physics, for example—lend themselves to controlled experimentation, and others, like astrophysics, to observation of natural phenomena. Many modern sciences have been able to aggressively pursue both, as technology has provided ever more powerful instruments, precisely designed and engineered materials, and suitably controlled environments. But in all fields the data gathered in these inquiries contribute to theories that describe, explain, and—in most cases—predict. These theories are then subjected to subsequent experimentation or observation—to be confirmed, refined, or eventually overturned—and the process evolves to more explanatory and definitive theories, often based on what are called natural laws, more specifically, the laws of physics.

Larry Smarr, a professor in the physics and astronomy departments at the University of Illinois at Champaign-Urbana, has traveled along the paths of both theory and experimentation. He organized the Frontiers symposium's session on computation, which illustrated the premise that computational science is not merely a new field or discipline or tool, but rather an altogether new and distinct methodology that has had a transforming effect on modern science: first, to permanently alter how scientists work, experiment, and theorize; and second, to expand their reach beyond the inherent limitations of the two other venerable approaches (Box 7.1).

Modern supercomputers, explained Smarr, "are making this alternative approach quite practical (Box 7.2). The result is a revolu-

BOX 7.1　COMPUTATION

The new terrain to be explored by computation is not a recent discovery, but rather comes from a recognition of the limits of calculus as the definitive language of science. Ever since the publication of Newton's *Philosophiae naturalis principia mathematica* in 1687, "calculus has been instrumental in the discovery of the laws of electromagnetism, gas and fluid dynamics, statistical mechanics, and general relativity. These classical laws of nature have been described," Smarr has written, by using calculus to solve partial differential equations (PDEs) for a continuum field (Smarr, 1985, p. 403). The power and also the limitation of these solutions, however, are the same: the approach is analytic. Where the equations are linear and separable, they can be reduced to the ordinary differential realm, and calculus and other techniques (like the Fourier transform), continued Smarr, "can give all solutions for the equations. However, progress in solving the *nonlinear* PDEs that govern a great portion of the phenomena of nature has been less rapid" (Smarr, 1985, p. 403).

Scientists refer to the results of their analysis as *the solution space*. Calculus approaches the solution of nonlinear or coupled differential equations analytically and also uses perturbation methods. These approaches have generally served the experimenter's purpose in focusing on a practical approximation to the full solution that—depending on the demands of the experiment—was usually satisfactory. But mathematicians down through history (Leonhard Euler, Joseph Lagrange, George Stokes, Georg Riemann, and Jules Henri Poincaré, to name only the most prominent) have not been as concerned with the practicalities of laboratory experiments as with the manifold complexity of nature. Smarr invoked mathematician Garrett Birkhoff's roll call of these men as predecessors to mathematician and computational pioneer John von Neumann, whose "main point was that mathematicians had nearly exhausted *analytical* methods, which apply mainly to *linear* differential equations and *special geometries*" (Smarr, 1985, p. 403). Smarr believes von Neumann "occupies a position similar to that of Newton" in the history of science because he realized the profound potential of computation to expand science's power to explore the full solution space; as Birkhoff put it, "to substitute *numerical* for analytical methods, tackling *nonlinear* problems in *general* geometries" (p. 403).

These distinctions are not mere mathematical niceties of scant concern to working scientists. The recognition that a vaster solution space may contain phenomena of crucial relevance to working physicists—and that computers are the only tools to explore it—has revitalized the science of dynamical systems and brought the word *chaos* into the scientific lexicon. Smarr quoted another mathematician, James Glimm, on their significance: "'Computers will affect science and technology at least as profoundly as did the invention of calculus. The reasons are the same. As with calculus, computers have increased and will increase enormously the range of solvable problems'" (Smarr, 1985, p. 403).

BOX 7.2   THE RIGHT HAND OF THE MODERN SCIENTIST

Since the late 1940s, Smarr noted, "computers have emerged as universal devices for aiding scientific inquiry. They are used to control laboratory experiments, write scientific papers, solve equations, and store data." But "the supercomputer has been achieving something much more profound; it has been transforming the basic methods of scientific inquiry themselves" (Smarr, 1992, p. 155).

"Supercomputer," Smarr explained, is a generic term, referring always to "the fastest, largest-memory machines made at a given time. Thus, the definition is a relative one. In 1990, to rate as a supercomputer, a machine needed to have about 1 billion bytes of directly addressable memory, and had to be capable of sustaining computations in excess of a billion floating point operations per second" (p. 155). These fairly common measures for rating computer power refer to an information unit (a byte contains eight of the basic binary units, 1 or 0, called bits) and the speed with which additions, subtractions, multiplications, or divisions of decimal numbers can be accomplished. These machines had between 100 and 1000 times the power of the personal computers commonly in use, and might cost in the neighborhood of $30 million. As he surveyed the growth of computational science, however, Smarr predicted that the "rapidly increasing speed and connectivity of networks will contribute to altering how humans work with supercomputers" (p. 162). He envisions that scientists and others working from personal and desktop machines will enjoy almost instantaneous access to a number of supercomputers around the country at once, with various parts of their programs running on each. He foresees that "the development of new software will make it seem as if they were occurring on one computer. In this sense the national network of computers will appear to be one giant 'metacomputer' that is as easy to use as today's most user-friendly personal machines" (p. 163).

Smarr credited von Neumann with the recognition, in the 1940s, that the technology of electronic digital computers "could be used to compute realistically complex solutions to the laws of physics and then to interrogate these solutions by changing some of the variables, as if one were performing actual experiments on physical reality" (p. 158). The American government's experience in World War II uncovered a number of practical applications of this principle and other uses for the computer's speed, "to aid in nuclear weapons design, code breaking, and other tasks of national security" (p. 157), reported Smarr. This effort developed in the succeeding decades into government centers with major laboratories—as reliant on their computers as on the electricity that powered them—at Los Alamos, New Mexico, and Livermore, California. "From the efforts of the large teams of scientists, engineers, and programmers who worked on the

continued on next page

BOX 7.2—*continued*

earlier generations of supercomputers, a methodology arose that is now termed computational science and engineering," according to Smarr (p. 157). It was their early high cost and the fact that only a few centers were feasible that made the government the primary purchaser of computers in the early years. But "throughout the 1980s, their use spread rapidly to the aerospace, energy, chemical, electronics, and pharmaceutical industries," summarized Smarr, "and by the end of the decade, more than 200 supercomputers were serving corporations" (p. 159).

Academic researchers and scientists joined this parade in ever greater legions during the 1980s, and support by the National Science Foundation led to the creation of five major supercomputer centers. Smarr, the director of the National Center for Supercomputing Applications (NCSA) at the University of Illinois, explained that the information infrastructure—built both of the machines themselves and the networks that connect them—is evolving to make "it possible for researchers to tap the power of supercomputers remotely from desktop computers in their own offices and labs" (p. 160). The national system called Internet, for example, has "encompassed all research universities and many four-year colleges, federal agencies and laboratories," and, Smarr reported, has been estimated to involve "more than a million users, working on hundreds of thousands of desktop computers," communicating with each other (p. 160).

SOURCE: Smarr (1992).

tion in our understanding of the complexity and variety inherent in the laws of nature" (Smarr, 1985, p. 403). Perhaps even more revolutionary, suggested Smarr, is what scientists can do with this new understanding: how they can use these more realistic solutions to engage in "a more constructive interplay between theory and experiment or observation than has heretofore been possible" (p. 403). The increasing power and economy of individual PCs or scientific workstations and their access through networks like Internet to supercomputers, predicted Smarr, will lead to a new way of doing daily science.

Imaging is one of the fundamental cornerstones of this new way of doing science. "A single color image," said Smarr, can represent "hundreds of thousands to millions of individual numbers" (Smarr, 1992, p. 161). Today's typical desktop computers "can even run animations directly on the screen. This capability, called scientific visualization, is radically altering the relationship between humans and the supercomputer" (p. 161). Indeed, the value of such pictures cannot be overestimated. Not only do they allow the compression of millions of data points into a coherent image, but that image is also a summary that appeals to the way the human brain processes information, because it establishes a picture and also moves that coherent picture through time, providing insight into the dynamics of a system or experiment that might well be obscured by a simple ream of numbers.

Among the participants in the session on computation, William Press, an astrophysicist at Harvard University, noted that starting several years ago, interdisciplinary conferences began to be awash with "beautiful movies" constructed to model and to help investigators visualize basic phenomena. "What is striking today," Press commented, "is that computer visualization has joined the mainstream of science. People show computer movies simply because you really cannot see what is going on without them." In fact Press believes strongly that computer modeling and visualization have become "an integral part of the science lying underneath."

The computation session's participants—also including Alan Huang of AT&T Bell Laboratories, Stephen Wolfram of Wolfram Research, and Jean Taylor from the Department of Mathematics at Rutgers University, discussed scientific computation from several diverse points of view. They talked about hardware and software, and about how both are evolving to enhance scientists' power to express their algorithms and to explore realms of data and modeling hitherto unmanageable.

THE MACHINERY OF THE COMPUTER

Digital Electronics

In the 1850s British mathematician George Boole laid the foundation for what was to become digital electronics by developing a system of symbolic logic that reduces virtually any problem to a series of true or false propositions. Since the output of his system was so elemental, he rejected the base 10 system used by most of the civilized world for centuries (most likely derived from ancient peoples counting on their fingers). The base 2 system requires a longer string of only two distinct symbols to represent a number (110101, for example, instead of 53). Since there are but two possible outcomes at each step of Boole's system, the currency of binary systems is a perfect match: an output of 1 signifies truth; a 0 means false. When translated into a circuit, as conceptualized by Massachusetts Institute of Technology graduate student Claude Shannon in his seminal paper "A Symbolic Analysis of Relay and Switching Circuits" (Shannon, 1938), this either/or structure can refer to the position of an electrical switch: either on or off.

As implemented in the computer, Boolean logic poses questions that can be answered by applying a series of logical constructs or steps called operators. The three operators found in digital computers are referred to by circuit designers as logic gates. These gates are physically laid down on boards and become part of the system's architecture. The computer's input is then moved through these gates as a series of bits. When a pair of bits encounters a gate, the system's elementary analysis occurs. The AND gate outputs 1 when both elements are 1, meaning, "It is true that the concept of *and* applies to the truth of both elements of the set"; conversely, the AND gate produces 0 as an output, meaning false, if either (or both) of the two elements in question is false; the OR gate outputs a 1, meaning true, if either element (or both) is a 1; with the NOT gate, the output is simply the reverse of whichever symbol (1 or 0) is presented.

Thus the physical structure of the basic elements of a circuit— including its gate design—becomes integral to analyzing the relationship of any pair of bits presented to it. When the results of this analysis occurring throughout the circuit are strung together, the final outcome is as simple as the outcome of any of its individual steps—on or off, 1 or 0, true or false—but that outcome could represent the answer to a complex question that had been reduced to a series of steps predicated on classical logic and using the three fundamental operators.

What is gained by reducing the constituent elements of the system to this simplest of all currencies—1 or 0, on or off—is sacrificed, however, in computational efficiency. Even the most straightforward of problems, when translated into Boolean logic and the appropriate series of logic gates, requires a great many such logical steps. As the computer age enters its fifth generation in as many decades, the elegant simplicity of having just two elemental outcomes strung together in logic chains begins to bump up against the inherent limitation of its conceptual framework. The history of computer programs designed to compete with human chess players is illustrative: as long as the computer must limit its "conceptual vision" to considering all of the alternatives in a straightforward fashion without the perspectives of judgment and context that human brains use, the chess landscape is just too immense, and the computer cannot keep up with expert players. Nonetheless, the serial computer will always have a role, and a major one, in addressing problems that are reducible to such a logical analysis. But for designers of serial digital electronics, the only way to improve is to run through the myriad sequences of simple operations faster. This effort turns out to reveal another inherent limitation, the physical properties of the materials used to construct the circuits themselves.

The Physical Character of Computer Circuitry

Integrated circuits (ICs), now basic to all computers, grew in response to the same generic challenges—moving more data faster and more efficiently—that are impelling scientists like Huang "to try to build a computer based on optics rather than electronics." By today's sophisticated standards, first-generation electronic computers were mastodons: prodigious in size and weight, and hungry for vast amounts of electricity—first to power their cumbersome and fragile vacuum tubes and then to cool and dissipate the excess heat they generated. The transistor in 1947 and the silicon chip on which integrated circuits were built about a decade later enabled development of the second- and third-generation machines. Transistors and other components were miniaturized and mounted on small boards that were then plugged into place, reducing complex wiring schemes and conserving space.

By the 1970s engineers were refining the IC to such a level of sophistication that, although no fundamental change in power and circuitry was seen, an entirely new and profound fourth generation of computers was built. ICs that once contained 100 or fewer components had grown through medium- and large-scale sizes to the very large-scale integrated (VLSI) circuit with as at least 10,000 components on a single chip, although some have been built with 4 million or more components. The physics involved in these small but complex circuits revolves around the silicon semiconductor.

The electrical conductivity of semiconductors falls between that of insulators like glass, which do not conduct electricity at all, and copper, which conducts it very well. In the period just after World War II, John Bardeen, Walter Brattain, and William Shockley were studying the properties of semiconductors. At that time, the only easy way to control the flow of electricity electronically (to switch it on or off, for example) was with vacuum tubes. But vacuum tube switches were bulky, not very fast, and required a filament that was heated to a high temperature in order to emit electrons that carried a current through the vacuum. The filament generated a great deal of heat and consumed large amounts of power. Early computers based on vacuum tubes were huge and required massive refrigeration systems to keep them from melting down. The Bardeen-Brattain-Shockley team's goal was to make a solid-state semiconductor device that could replace vacuum tubes. They hoped that it would be very small, operate at room temperature, and would have no power-hungry filament. They succeeded, and the result was dubbed the transistor. All modern computers are based on fast transistor switches. A decade later, the team was awarded a Nobel Prize for their accomplishment, which was to transform modern society.

How to build a chip containing an IC with hundreds of thousands of transistors (the switches), resistors (the brakes), and capacitors (the electricity storage tanks) wired together into a physical space about one-seventh the size of a dime was the question; photolithography was the answer. Literally "writing with light on stone," this process allowed engineers to construct semiconductors with ever finer precision. Successive layers of circuitry can now be laid down in a sort of insulated sandwich that may be over a dozen layers thick.

As astounding as this advance might seem to Shannon, Shockley, and others who pioneered the computer only half a lifetime ago, modern physicists and communications engineers continue to press the speed envelope. Huang views traditional computers as fundamentally "constrained by inherent communication limits. The fastest transistors switch in 5 picoseconds, whereas the fastest computer runs with a 5 nanosecond clock." This difference of three orders of magnitude beckons, suggesting that yet another generation of computers could be aborning. Ablinking, actually: in optical computers a beam of light switched on represents a 1, or true, signal; switching it off represents a 0, or false, signal.

NEW DIRECTIONS IN HARDWARE FOR COMPUTATION

Computing with Light

The major application of optics technology thus far has been the fiber-optic telecommunications network girding the globe. Light rays turn out to be superior in sending telephone signals great distances because, as Huang described it, with electronic communication, "the farther you go, the greater the energy required, whereas photons, once launched, will travel a great distance with no additional energy input. You get the distance for free. Present engineering has encountered a crossover point at about 200 microns, below which distance it is more efficient to communicate with electrons rather than photons. This critical limit at first deflected computer designers' interest in optics, since rarely do computer signals travel such a distance in order to accomplish their goal." To designers of telecommunications networks, however, who were transmitting signals on the order of kilometers, the "extra distance for free" incentive was compelling. The situation has now changed with the advent of parallel processing, since a large portion of the information must move more than 200 microns. Energy is not the only consideration, however. As the physical constraints inherent in electronic transmission put a cap on the speed at which the circuitry can connect, optical digital computing presents a natural alternative. "The next challenge for advanced optics," predicted Huang, "is in the realm of making connections between computer chips, and between individual gates. This accomplished, computers will then be as worthy of the designation optical as they are electronic."

Huang was cautious, however, about glib categorical distinctions between the two realms of optics and electronics. "How do you know when you really have an optical computer?" he asked. "If you took out all the wires and replaced them with optical connections, you would still have merely an optical version of an electronic computer. Nothing would really have been redesigned in any fundamental or creative way." He reminded the audience at the Frontiers symposium of the physical resemblance of the early cars built at the turn of the century to the horse-drawn carriages they were in the process of displacing and then continued, "Suppose I had been around then and come up with a jet engine? Would the result have been a jet-powered buggy? If you manage to come up with a jet engine, you're better off putting wings on it and flying it than trying to adapt it to the horse and buggy." A pitfall in trying to retrofit one technology onto another, he suggested, is that "it doesn't work"—it is necessary instead to redesign "from conceptual scratch." But the process is a long and involved one. The current state of the art of fully optical machines, he remarked wryly, is represented by one that "is almost smart enough to control a washing machine but is a funky-looking thing that will no doubt one day be regarded as a relic alongside the early mechanical computers with their heavy gears and levers and hand-cranks" (Figure 7.1).

The Optical Environment

Optics is more than just a faster pitch of the same basic signal, although it can be indisputably more powerful by several orders of magnitude. Huang surveyed what might be called the evolution of optical logic:

FIGURE 7.1 The digital optical processor. Although currently only about as powerful as a silicon chip that controls a home appliance such as a dishwasher, the optical processor may eventually enable computers to operate 1000 times faster than their electronic counterparts. It demonstrates the viability of digital optical processing technology. (Courtesy of A. Huang.)

> I believe electronics—in terms of trying to get the basic stuff of transmission in and out of the chips—will reach an inherent speed limit at about 1 gigahertz. The next stage will involve subdividing the chip differently to use more optical modulators or microlasers. A clever rearrangement of the design should enable getting the systems to work at about 10 gigahertz, constrained only by the limits inherent to the integrated circuit wiring on the chip itself. Surmounting this limit will require building logic gates to connect one array to another, with each logic gate using an optical modulator. Such an architecture should enable pushing up to the limits of the semiconductor, or about 100 gigahertz.

Not only a pioneer in trying to develop this new optical species of computer, Huang has also been cited with teaching awards for his thoughtful communicative style in describing it. He worries that the obsession with measuring speed will obscure the truly revolutionary nature of optical computing—the process, after all, is inherently constrained not only by the physics of electronic transmission, but also by a greater limit, as Einstein proved, the speed of light. But photons of light moving through a fiber tunnel—in addition to being several orders of magnitude faster than electrons moving through a wire on a chip—possess two other inherent advantages that far outweigh their increased speed: higher bandwidth and greater connectivity.

In order to convey the significance of higher bandwidth and greater connectivity, Huang presented his "island of Manhattan" metaphor:

> Everybody knows that traffic in and out of Manhattan is really bottlenecked. There are only a limited number of tunnels and bridges, and each of them has only a limited number of lanes. Suppose we use the traffic to represent information. The number of bridges and tunnels would then represent the connectivity, and the number of traffic lanes would represent the bandwidth.
>
> The use of an optical fiber to carry the information would correspond to building a tunnel with a hundred times more lanes. Transmitting the entire *Encyclopedia Britannica* through a wire would take 10 minutes, while it would take only about a second with optical fiber.
>
> Instead of using just an optical fiber, the systems we are working on use lenses. Such an approach greatly increases the connectivity since each lens can carry the equivalent of hundreds of optical fibers. This would correspond to adding hundreds of more tunnels and bridges to Manhattan.
>
> This great increase in bandwidth and connectivity would greatly ease the flow of traffic or information. Try to imagine all the telephone wires in the world strung together in one giant cable. All of the telephone calls can be carried by a series of lenses.

The connectivity of optics is one reason that optical computers may one day relegate their electronic predecessors to the historical junkyard (Figure 7.2). Virtually all computing these days is serial;

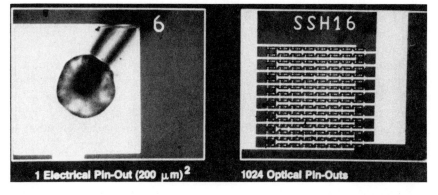

FIGURE 7.2 Optical vs. electrical pin-outs. The greater connectivity of optics is indicated in this comparison of a VLSI bonding pad with an array of SEED optical logic gates in the equivalent amount of area. (Courtesy of A. Huang.)

the task is really to produce one output, and the path through the machine's circuitry is essentially predetermined, a function of how to get from input to output in a linear fashion along the path of the logic inherent to, and determined by, the question. Parallel computing—carried out with a fundamentally different machine architecture that allows many distinct or related computations to be run simultaneously—turns this structure on its ear, suggesting the possibility of many outputs, and also the difficult but revolutionary concept of physically modifying the path as the question is pursued in the machine. The science of artificial intelligence has moved in this direction, inspired by the manifest success of massively parallel "wiring" in human and animal brains.

With the connectivity advantage outlined, Huang extended his traffic metaphor: "Imagine if, in addition to sending so many more cars through, say, the Holland Tunnel, the cars could actually pass physically right through one another. There would be no need for stoplights, and the present nightmare of traffic control would become trivial." With the pure laser light generally used in optics, beams can pass through one another with no interference, regardless of their wavelength. By contrast, Huang pointed out, "In electronics, you're always watching out for interference. You can't send two electronic signals through each other."

Although the medium of laser light offers many advantages, the physical problems of manipulating it and moving it through the logic gates of ever smaller optical architectures present other challenges. In responding to them, engineers and designers have developed a remarkable technique, molecular beam epitaxy (MBE), which Huang has used to construct an optical chip of logic gates, lenses, and mirrors atom by atom, and then layer by layer (Box 7.3; Figure 7.3). It is possible, he explained, to "literally specify . . . 14 atoms of this, 37 of that, and so on, for a few thousand layers." Because of the precision of components designed at the atomic level, the light itself can be purified into a very precise laser beam that will travel through a hologram of a lens as predictably as it would travel through the lens itself. This notion brings Huang full circle back to his emphasis on deflecting simple distinctions between electronics and optics: "With MBE technology, we can really blur the distinction between what is optical, what is electronic, and what is solid-state physics. We can actually crunch them all together, and integrate things on the atomic level." He summarized, "All optical communication is really photonic, using electromagnetic waves to propagate the signal, and all interaction is really electronic, since electrons are always involved at

BOX 7.3 BUILDING THE OPTICAL LENS

While the medium of laser light offers so many advantages, the physical problem of manipulating and moving it through the logic gates of ever smaller optical architecture presents challenges. "In optical systems, the basic pipeline architecture to move the data consists of various lens arrangements, then a plane of optical logic gates, another array of lenses, another array of logic gates," up to the demands of the designer, said Alan Huang. "We discovered that the mechanics of fabrication—trying to shrink the size of the pipeline—poses a real problem," he continued, "prompting us to search for a new way to build optics. We've arrived at the point where we grow most of our own devices with molecular beam epitaxy (MBE)."

Designers and engineers using MBE technology start with gallium arsenide as a conductor for the optical chip because electrons move through it faster than through silicon and it is inherently more capable of generating and detecting light. The process involves actually fabricating layer after layer of atomic material onto the surface of the wafer. Elements of aluminum, gallium, and other fabricating materials are superheated, and then a shutter for a particular element is opened onto the vacuum chamber in which the wafer has been placed. The heated atoms migrate into the vacuum, and by what Huang described as "the popcorn effect" eventually settle down on the surface in a one-atom-thick lattice, with their electrons bonding in the usual way. "So you just build the lattice, atom by atom, and then layer by layer. What's remarkable about it is, in a 2-inch crystal, you can see plus or minus one atomic layer of precision: I can literally specify—give me 14 atoms of this, 37 thick of that, for a few thousand layers," Huang explained.

But there is an even more astounding benefit to the optical lens, involving lasers. The MBE process permits such precision that a virtual laser-generating hologram of the lens will work as well as the lens itself. Thus, though Huang may pay an outside "grower" $4000 to build a gallium arsenide chip with lenses on it, from that one chip 10 million lasers can then be stamped out by photolithography. Normal, physical lenses are generally required to filter the many wavelengths and impurities in white light. With pure laser light of a precise frequency, the single requirement of deflecting the beam a certain angle can be accomplished by building a quantum well sensitive to that frequency, and then collapsing the whole structure into something physicists call a Fresnel zone plate.

The prospect of this profusion of lasers, Huang suggested, should surprise no student of the history of technology: "You now probably have more computers in your house than electric motors. Soon, you will have more lasers than computers. It kind of sneaks up on you."

FIGURE 7.3 The extraordinary precision of this molecular beam epitaxy machine allows scientists to build semiconductor chips atomic layer by atomic layer. (Courtesy of A. Huang.)

some level." Extraordinary advances in engineering thus permit scientists to reexamine some of their basic assumptions.

THE COMPUTER'S PROGRAMS—
ALGORITHMS AND SOFTWARE

Algorithms—the term and the concept—have been an important part of working science for centuries, but as the computer's role grows, the algorithm assumes the status of a vital methodology on which much of the science of computation depends (Harel, 1987). Referring to the crucial currency of computation, speed, Press stated that "there have been at least several orders of magnitude of computing speed gained in many fields due to the development of new algorithms."

An algorithm is a program compiled for people, whereas software is a program compiled for computers. The distinction is sometimes meaningless, sometimes vital. In each case, the goal of solving a particular problem drives the form of the program's instructions. A successful algorithm in mathematics or applied science provides definitive instructions for an unknown colleague to accomplish a particular task, self-sufficiently and without confusion or error. When that algorithm involves a computer solution, it must be translated through the appropriate program and software. But when around 300 B.C. Euclid created the first nontrivial algorithm to describe how to find the greatest common divisor of two positive integers, he was providing for all time a definitive recipe for solving a generic problem. Given the nature of numbers, Euclid's algorithm is eternal and will continue to be translated into the appropriate human language for mathematicians forever.

Not all algorithms have such an eternal life, for the problems whose solutions they provide may themselves evolve into a new form, and new tools, concepts, and machinery may be developed to address them. Press compiled a short catalog of recent "hot" algorithms in science and asked the symposium audience to refer to their own working experience to decide whether the list represented a "great edifice or a junkpile." Included were fractals and chaos, simulated annealing, Walsh functions, the Hartley and fast Fourier transforms, fuzzy sets, and catastrophe theory. The most recent among these, which Press speculated could eventually rival the fast Fourier transform (FFT) as a vital working tool of science, is the concept of wavelets.

Over the last two decades, the Fourier transform (FT) has reduced by hundreds of thousands of hours the computation time working scientists would otherwise have required for their analyses. Although it was elucidated by French mathematician J.B.J. Fourier "200 years ago," said Press, "at the dawn of the computer age it was impractical for people to use it in trying to do numerical work" until J.W. Cooley and J.W. Tukey devised the FFT algorithm in the 1960s. Simply stated, the FFT makes readily possible computations that are a factor of 10^6 more complicated than would be computable without it. Like any transform process in mathematics, the FT is used to simplify or speed up the solution of a particular problem. Presented with a mathematical object X, which usually in the world of experimental science is a time series of data points, the FT first changes, or transforms, X into a different but related mathematical object X'.

The purpose of the translation is to render the desired (or a related) computation more efficient to accomplish. The FT has been found useful for a vast range of problems. Press is one of the authors of *Numerical Recipes: the Art of Scientific Computing* (Press et al., 1989), a text on numerical computation that explains that "a physical process can be described either in the *time domain*, or else in the *frequency domain*. For many purposes it is useful to think of these as being two different representations of the same function. One goes back and forth between these two representations by means of the *Fourier transform* equations" (p. 381). The FT, by applying selective time and frequency analyses to an enormous amount of raw data, extracts the essential waveform information into a smaller data set that nevertheless has sufficient information for the experimenter's purposes.

Wavelets

The data signals presented to the computer from certain phenomena reflect sharp irregularities when sudden or dramatic transitions occur; the mathematical power of the FT, which relies on the creation and manipulation of trigonometric sines and cosines, does not work well at these critical transition points. For curves resulting from mathematical functions that are very complex and that display these striking discontinuities, the wavelet transform provides a more penetrating analytic tool than does the FT or any other mathematical operation. "Sine waves have no localization," explained Press, and thus do not bunch up or show a higher concentration at data points where the signal is particularly rich with detail and change. The power of wavelets comes in part from their variability of scale, which permits the user to select a wavelet family whose inherent power to resolve fine details matches the target data.

The fundamental operation involves establishing a banded matrix for the given wavelet family chosen. These families are predicated on the application of a specific set of coefficient numbers—which were named for Ingrid Daubechies, a French mathematician whose contributions to the theory have been seminal—and are called Daub 4 or Daub 12 (or Daub X), with a greater number of coefficients reflecting finer desired resolution. Press demonstrated to the symposium's audience how the Daub 4 set evolves, and he ran through a quick summary of how a mathematician actually computes the wavelet transform. The four special coefficients for Daub 4 are as follows:

$$c_0 = (1 + \sqrt{3}) / 4\sqrt{2} \quad c_1 = (3 + \sqrt{3}) / 4\sqrt{2}$$
$$c_2 = (3 - \sqrt{3}) / 4\sqrt{2} \quad c_3 = (1 - \sqrt{3}) / 4\sqrt{2}$$

The banded matrix is established as follows: install the numbers c_0 through c_3 at the upper left corner in the first row, followed by zeros to complete the row; the next row has the same four numbers directly below those on the first row, but with their order reversed and the c_2 and c_0 terms negative; the next row again registers the numbers in their regular sequence, but displaced two positions to the right, and so on, with zero occupying all other positions. At the end of a row of the matrix, the sequence wraps around in a circulant.

The mathematical character that confers on the wavelet transform its penetrating power of discrimination comes from two properties: first, the inverse of the matrix is found by transposing it—that is, the matrix is what mathematicians call orthogonal; second, when the fairly simple process of applying the matrix and creating a table of sums and differences is performed on a data set, a great many vanishing moments result—contributions so small that they can be ignored with little consequence. In practical terms, a mathematical lens seeks out smooth sectors of the data, and confers small numbers on them to reduce their significance relative to that of rough sectors, where larger numbers indicate greater intricacy or sudden transition. Press summarized, "The whole procedure is numerically stable and very fast, faster than the fast Fourier transform. What is really interesting is that the procedure is hierarchical in scale, and thus the power of wavelets is that smaller components can be neglected; you can just throw them out and compress the information in the function. That's the trick."

Press dramatically demonstrated the process by applying the wavelet transform to a photograph that had been scanned and digitized for the computer (Figure 7.4). "I feel a little like an ax murderer," he commented, "but I analyzed this poor woman into wavelets, and I'm taking her apart by deleting those wavelets with the smallest coefficients." This procedure he characterized as fairly rough: a sophisticated signal processor would actually run another level of analysis and assign a bit code to the coefficient numbers. But with Press simply deleting the smallest coefficients, the results were striking. With 77 percent of the numbers removed, the resulting photograph was almost indistinguishable from the original. He went to the extreme of removing 95 percent of the signal's content using the wavelets to retain "the information where the contrast is large," and still reconstructed a photograph unmistakably comparable to the original. "By doing good signal processing and assigning bits rather than just deleting the smallest ones, it is possible to make the 5 percent picture look about as good as the original photograph," Press explained.

Press pointed out that early insights into the wavelet phenomenon came from work on quadrature filter mirrors, and it is even conceivable that they may help to better resolve the compromised pictures coming back from the Hubble Space Telescope. And while image enhancement and compression may be the most accessible applications right now, wavelets could have other powerful signal-processing uses wherever data come by way of irregular waveforms, such as for speech recognition, in processing of the sound waves used in geological explorations, in the storage of graphic images, and

A

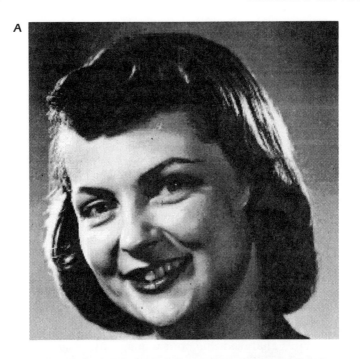

B

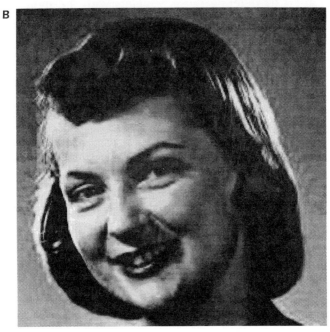

C

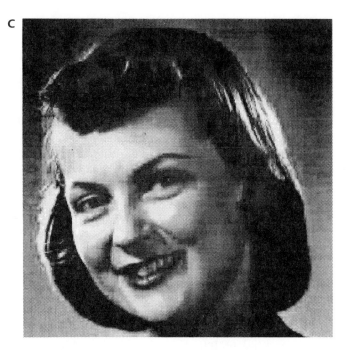

FIGURE 7.4 Application of a wavelet transform to a photograph scanned and digitized for the computer. (A) Original photograph. (B) Reconstructed using only the most significant 23 percent of wavelets. (C) Reconstructed using only the most significant 5 percent of wavelets, that is, discarding 95 percent of the original image. (Courtesy of W. Press.)

in faster magnetic resonance imaging and other scanners. Work is even being done on a device that would monitor the operating sounds of complex equipment, such as jet engines or nuclear reactors, for telltale changes that might indicate an imminent failure. These and many other problems are limited by the apparent intractability of the data. Transforming the data into a more pliable and revelatory form is crucial, said Press, who continued, "A very large range of problems—linear problems that are currently not feasible—may employ the numerical algebra of wavelet basis matrices and be resolved into sparse matrix problems, solutions for which are then feasible."

Experimental Mathematics

The work of Jean Taylor, a mathematician from Rutgers University, illustrates another example of the computer's potential to both support research in pure mathematics and also to promote more general understanding. Taylor studies the shapes of surfaces that are stationary, or that evolve in time, looking for insights into how energies control shapes; the fruits of her work could have very direct applications for materials scientists. But the level at which she approaches these problems is highly theoretical, very complex, and specialized. She sees it as an outgrowth of work done by mathematician Jesse Douglas, who won one of the first Fields medals for his work on minimal surfaces. She illustrated one value of the computer by explaining to the symposium the dilemma she often faces when trying to respond to queries about what she does. She may begin to talk about minimal surfaces at their extremes, which turn into polyhedral problems; this suggests to her listeners that she does discrete and computational geometry, which, she says, as it is normally defined does not describe her approach. Work on minimal surfaces is usually classified as differential geometry, but her work seems to fall more into the nondifferential realm. Even trained geometers often resort to categorizing her in not-quite-accurate pigeonholes. "It is much easier to show them, visually," said Taylor, who did just that for the symposium's audience with a pair of videos rich with graphical computer models of her work, and the theory behind it.

"Why do I compute?" asked Taylor. "I have developed an interaction with the program that makes it an instrumental part of my work. I can prove to others—but mainly to myself—that I really understand a construction by being able to program it." She is in the business of developing theorems about the equilibrium and growth shapes of crystals. The surfaces she studies often become "singular, producing situations where the more classical motion by curvature breaks down," she emphasized. No highly mathematical theory alone can render such visual phenomena with the power and cogency of a picture made from evolving the shape with her original computer model (Figure 7.5).

This sophisticated dialectic between theory and computer simulation also "suggests other theorems about other surface tension functions," continued Taylor, who clearly is a theoretical mathematician running experiments on her computer, and whose results often go far beyond merely confirming her conjectures.

Stephen Wolfram is ideally positioned to evaluate this trend. "In the past," he told symposium participants, "it was typically the case

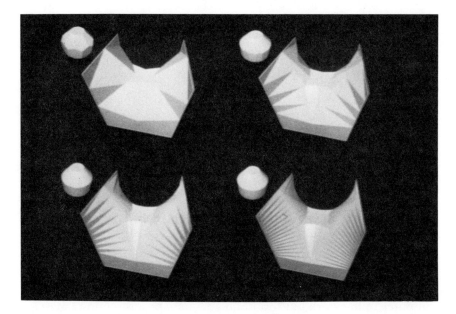

FIGURE 7.5 Four surfaces computed by Taylor, each having the same eight line segments (which lie on the edges of a cube) as a boundary, but with a different surface energy function for each. Each surface energy function is specified by the equilibrium single-crystal shape (the analog of a single soap-bubble) at the upper left of its surface. Computations such as this provide not only the raw material for making conjectures but also a potential method for proving such conjectures, once the appropriate convergence theorems are proved. (Courtesy of J. Taylor.)

that almost all mathematicians were theoreticians, but during the last few years there has arisen a significant cohort of mathematicians who are not theoreticians but instead are experimentalists. As in every other area of present day science, I believe that before too many years have passed, mathematics will actually have more experimentalists than theorists."

In his incisive survey of the computer and the changing face of science, *The Dreams of Reason*, physicist Heinz Pagels listed the rise of experimental mathematics among the central themes in what he considered "a new synthesis of knowledge based in some general way on the notion of complexity. . . . The material force behind this change is the computer" (Pagels, 1988, p. 36). Pagels maintained that the evolution concerns the very order of knowledge. Before the rise of empirical science, he said, the "architectonic of the natural sciences (natural philosophy), in accord with Aristotelian canons, was established by the *logical* relation of one science or another" (p. 39). In the next era, the instrumentation (such as the microscope) on which empirical science relied, promoted reductionism: "the properties of the small things determined the behavior of the larger things," according to Pagels (p. 40), who asserted that with the computer's arrival "we may begin to see the relation between various sciences in entirely new dimensions" (p. 40). Pagels believed that the common tool of the computer, and the novel perspective it provides on the world, are provoking a horizontal integration among the sciences, necessary in order to "restructure our picture of reality" (p. 42).

Implementing Software for Scientists

One of those providing a spotlight on this new picture of scientific reality is Wolfram, who runs Wolfram Research, Inc., and founded the Center for Complex Systems at the Urbana campus of the University of Illinois. Wolfram early realized the potential value of the computer to his work but found no adequate programming language that would fully utilize its power. He began working with a language called SMP and has since created and continued to refine a language of his own, which his company now markets as the software package *Mathematica*, used by some 200,000 scientists all over the world. He surveyed the Frontiers audience and found half of its members among his users.

Mathematica is a general system for doing mathematical computation and other applications that Press called "a startlingly good tool." As such, it facilitates both numerical and symbolic computation, develops elaborate sound and graphics to demonstrate its results, and provides "a way of specifying algorithms" with a programming language that Wolfram hopes may one day be accepted as a primary new language in the scientific world, although he pointed out how slowly such conventions are established and adopted (Fortran from the 1960s and C from the early 1970s being those most widely used now). Wolfram expressed his excitement about the prospects to "use that sort of high-level programming language to represent models and ideas in science. With such high-level languages, it becomes much more realistic to represent an increasing collection of the sorts of models that one has, not in terms of traditional algebraic formulae but instead in terms of algorithms, for which *Mathematica* provides a nice compact notation."

Similarly, noted Wolfram, "One of the things that has happened as a consequence of *Mathematica* is a change in the way that at least some people teach calculus. It doesn't make a lot of sense anymore to focus a calculus course around the evaluation of particular integrals because the machine can do it quite well." One of the mathematicians in the audience, Steven Krantz from the session on dynamical systems, was wary of this trend, stating his belief that "the techniques of integration are among the basic moves of a mathematician—and of a scientist. He has got to know these things." From his point of view, Wolfram sees things differently. "I don't think the mechanics of understanding how to do integration are really relevant or important to the intellectual activity of science," he stated. He has found that, in certain curricula, *Mathematica* has allowed teachers who grasp it a freedom from the mechanics to focus more on the underlying ideas.

THE SCIENCES OF COMPLEXITY

Cellular Automata

Wolfram, according to Pagels, "has also been at the forefront of the sciences of complexity" (Pagels, 1988, p. 99). Wolfram is careful to distinguish the various themes and developments embraced by Pagels' term—which "somehow relate to complex systems"—from a truly coherent science of complexity, which he concedes at present is only "under construction." At work on a book that may establish the basis for such a coherent view, Wolfram is a pioneer in one of the most intriguing applications of computation, the study of cellular automata. "Computation is emerging as a major new approach to science, supplementing the long-standing methodologies of theory and experiment," said Wolfram. Cellular automata, "cells" only in the sense that they are unitary and independent, constitute the mathematical building blocks of systems whose behavior may yield crucial insights about form, complexity, structure, and evolution. But they are not "real," they have no qualities except the pattern they grow into, and their profundities can be translated only through the process that created them: computer experimentation.

The concept of cellular automata was devised to explore the hypothesis that "there is a set of mathematical mechanisms common to many systems that give rise to complicated behavior," Wolfram has explained, suggesting that the evolution of such chaotic or complex behavior "can best be studied in systems whose construction is as simple as possible" (Wolfram, 1984, p. 194). Cellular automata are nothing more than abstract entities arranged in one or two (conceivably and occasionally more) dimensions in a computer model, each one possessing "a value chosen from a small set of possibilities, often just 0 and 1. The values of all cells in the cellular automaton are simultaneously updated at each 'tick' of a clock according to a definite rule. The rule specifies the value of a cell, given its previous value and the values of its nearest neighbors or some other nearby set of cells" (p. 194).

What may seem at first glance an abstract and insular game turns out to generate remarkably prescient simulations that serve, according to Wolfram, "as explicit models for a wide variety of physical systems" (Wolfram, 1984, p. 194). The formation of snowflakes is one example. Beginning with a single cell, and with the rules set so that one state represents frozen and the other state water vapor, it can be shown that a given cell changes from a vapor state to a frozen state only when enough of its neighbors are vaporous and will not thereby inhibit the dissipation of enough heat to freeze. These rules conform to what we know about the physics of water at critical temperatures, and the resulting simulation based on these simple rules grows remarkably snowflake-like automata (Figure 7.6).

Cellular automata involve much more than an alternative to running complicated differential equations as a way of exploring snowflake and other natural structures that grow systematically; they may be viewed as analogous to the digital computer itself. "Most of [their] properties have in fact been conjectured on the basis of patterns generated in computer experiments" (Wolfram, 1984, p. 194). But their real power may come if they happen to fit the definition of a universal computer, a machine that can solve all computable problems. A computable problem is one "that can be solved in a finite time by following definite algorithms" (p. 197). Since the basic operation of a computer involves a myriad of simple binary changes, cellular automata—if the rules that govern their evolution are chosen appropriately—can *be* computers. And "since any physical process can be represented as a computational process, they can mimic the action of any physical system as well" (p. 198).

Resonant with the fundamental principles discussed in the session on dynamical systems, Wolfram's comments clarified the significance of these points: "Differential equations give adequate models for the overall properties of physical processes," such as chemical reactions, where the goal is to describe changes in the total concentration of molecules. But he reminded the audience that what they are really seeing, when the roughness of the equations is fully acknowledged, is actually an average change in molecular concentration based on countless, immeasurable random walks of individual molecules. But the old science *works*. At least the results produced a close enough fit, for many systems and the models based on differential equations that described them. "However," Wolfram continued,

FIGURE 7.6 Computer-generated cellular automaton pattern. The simulated snowflake grows outward in stages, with new cells added according to a simple mathematical rule. Hexagonal rule = 101010_2. (Reprinted with permission from Norman H. Packard and Stephen Wolfram. Copyright © 1984 by Norman H. Packard and Stephen Wolfram.)

"there are many physical processes for which no such average description seems possible. In such cases differential equations are not available and one must resort to direct simulation. The only feasible way to carry out such simulations is by computer experiment: essentially no analysis of the systems for which analysis is necessary could be made without the computer."

The Electronic Future Arrives

The face of science is changing. Traditional barriers between fields, based on specialized knowledge and turf battles, are falling. "The material force behind this change is the computer," emphasized Pagels (Pagels, 1988, p. 36), who demonstrated convincingly how the computer's role was so much more than that of mere implement.

Science still has the same mission, but the vision of scientists in this new age has been refined. Formerly, faced with clear evidence of complexity throughout the natural world, scientists had to limit the range of vision, or accept uneasy and compromised descriptions. Now, the computer—in Pagels' words the ultimate "instrument of complexity"—makes possible direct examination and experimentation by way of computation, the practice of which he believed might deserve to be classed as a distinct branch of science, alongside the theoretical and the experimental. When colleagues object that computer runs are not real experiments, the people that Pagels wrote about, and others among the participants in the Frontiers symposium, might reply that, in many experimental situations, computation is the *only* legitimate way to simulate and capture what is really happening.

Wolfram's prophecy may bear repetition: "As in every other area of present day science, I believe that before too many years have passed mathematics will actually have more experimentalists than theorists." One of the reasons for this must be that no other route to knowledge and understanding offers such promise, not just in mathematics but in many of the physical sciences as well. Pagels called it "the rise of the computational viewpoint of physical processes," reminding us that the world is full of dynamical systems that are, in essence, computers themselves: "The brain, the weather, the solar system, even quantum particles are all computers. They don't *look* like computers, of course, but what they are computing are the consequences of the laws of nature" (Pagels, 1988, p. 45). Press talked about the power of traditional algorithms to capture underlying laws; Pagels expressed a belief that "the laws of nature are algorithms that control the development of the system in time, just like real programs do for computers. For example, the planets, in moving around the sun, are doing analogue computations of the laws of Newton" (p. 45).

Indisputably, the computer provides a new way of seeing and modeling the natural world, which is what scientists do, in their efforts to figure out and explain its laws and principles. Computational science, believes Smarr, stands on the verge of a "golden age," where the exponential growth in computer speed and memory will not only continue (Box 7.4), but together with other innovations achieve something of a critical mass in fusing a new scientific world: "What is happening with most sciences is the transformation of science to a digital form. In the 1990s, a national information infrastructure to support digital science will arise, which will hook together supercomputers, massive data archives, observational and experimental instruments, and millions of desktop computers." We are well on the way, he believes, to "becoming an electronic scientific community" (Smarr, 1991, p. 101).

BOX 7.4 PARALLEL COMPUTING

Whether serial or parallel, explained neuroscientist James Bower from the California Institute of Technology, all computers "consist of three basic components: processors, memory, and communication channels," which provide a way to distinguish and compare them. Up until the 1970s, said Larry L. Smarr of the National Center for Supercomputing Applications (NCSA), computer designers intent on increasing the speed and power of their serial machines concentrated on "improving the microelectronics" of the single central processing unit (CPU) where most of the computation takes place, but they were operating with a fundamental constraint.

The serial CPUs are scalar uniprocessors, which means that they operate—no matter how fast—on but one number at a time. Smarr noted that Seymour Cray, called "the preeminent supercomputer designer of recent decades, successfully incorporated a [fundamental] improvement called vector processing," where long rows of numbers called vectors could be operated on at once. The Cray-1 model increased the speed of computing by an order of magnitude and displaced earlier IBM and Control Data Corporation versions (that had previously dominated the market in the 1960s and 1970s) as the primary supercomputer used for scientific and advanced applications. To capitalize on this design, Cray Research, Inc., developed succeeding versions, first the X-MP with four, and then the Y-MP with eight powerful, sophisticated vector processors able to "work in parallel," said Smarr. "They can either run different jobs at the same time or work on parts of a single job simultaneously." Such machines with relatively few, but more powerful, processors are classified as coarse-grained. They "are able to store individual programs at each node," and usually allow for very flexible approaches to a problem because "they operate in the Multiple-Instruction Multiple-Data (MIMD) mode," explained Bower.

By contrast, another example of parallel processing put to increasing use in the scientific environment is the Connection Machine, designed by W. Daniel Hillis, and manufactured by Thinking Machines Corporation. Like the biological brains that Hillis readily admits inspired its design, the Connection Machine's parallelism is classified as fine-grained. In contrast to the more powerful Cray machines, it has a large number of relatively simple processors (nodes), and operates on data stored locally at each node, in the Single-Instruction Multiple-Data (SIMD) mode.

Hillis and his colleagues realized that human brains were able to dramatically outperform the most sophisticated serial computers in visual processing tasks. They believed this superiority was inherent in the fine-grained design of the brain's circuitry. Instead of relying

continued on next page

BOX 7.4—*continued*

on one powerful central processing unit to search through all of the separate addresses where a bit memory might be stored in a serial computer, the Connection Machine employs up to 65,536 simpler processors (in its present incarnation), each with a small memory of its own. Any one processor can be assigned a distinct task—for example, to attend to a distinct point or pixel in an image—and they can all function at once, independently. And of particular value, any one processor can communicate with any other and modify its own information and task pursuant to such a communication. The pattern of their connections also facilitates much shorter information routes: tasks proceed through the machine essentially analyzing the best route as they proceed. The comparatively "blind" logic of serial machines possesses no such adaptable plasticity. "Each processor is much less powerful than a typical personal computer, but working in tandem they can execute several billion instructions per second, a rate that makes the Connection Machine one of the fastest computers ever constructed," wrote Hillis (1987, p. 108).

Speed, cost per component, and cost per calculation have been the currencies by which computers have been evaluated almost from their inception, and each successive generation has marked exponential improvement in all three standards. Nonetheless, Smarr believes the industry is approaching a new level of supercomputing unlike anything that has gone before: "By the end of the 20th century, price drops should enable the creation of massively parallel supercomputers capable of executing 10 trillion floating-point operations per second, 10,000 times the speed of 1990 supercomputers." This astounding change derives from the massively parallel architecture, which "relies on the mass market for ordinary microprocessors (used in everything from automobiles to personal computers to toasters) to drive down the price and increase the speed of general-design chips. Supercomputers in this new era will not be confined to a single piece of expensive hardware located in one spot, but rather will be built up by linking "hundreds to thousands of such chips together with a high-efficiency network," explained Smarr.

BIBLIOGRAPHY

Harel, David. 1987. Algorithmics: The Spirit of Computing. Addison-Wesley, Reading, Mass.

Hillis, W. Daniel. 1987. The connection machine. Scientific American 256(June):108-115.

Pagels, Heinz R. 1988. The Dreams of Reason: The Computer and the Rise of the Sciences of Complexity. Simon and Schuster, New York.

Press, William, Brian P. Flannery, Saul A. Teukolsky, and William T. Vetterling. 1989. Numerical Recipes: The Art of Scientific Computing. Cambridge University Press, New York.

Shannon, Claude E. 1938. A symbolic analysis of relay switching and controls. Transactions of the American Institute of Electrical Engineers 57:713-723.

Smarr, Larry L. 1985. An approach to complexity: Numerical computations. Science 228(4698):403-408.

Smarr, Larry L. 1991. Extraterrestrial computing: Exploring the universe with a supercomputer. Chapter 8 in Very Large Scale Computation in the 21st Century. Jill P. Mesirov (ed.). Society for Industrial and Applied Mathematics, Philadelphia.

Smarr, Larry L. 1992. How supercomputers are transforming science. Yearbook of Science and the Future. Encyclopedia Britannica, Inc., Chicago.

Wolfram, Stephen. 1984. Computer software in science and mathematics. Scientific American 251(September):188-203.

Wolfram, Stephen. 1991. Mathematica: A System for Doing Mathematics by Computer. Second edition. Addison-Wesley, New York.

8

Research and Regulation: Science's Contribution to the Public Debate

Regional and global air pollution problems have become highly sensitive and politicized subjects. Potential consequences to the biosphere, the health of its inhabitants, and the very future of the planet are the subjects of intense debate. Currently, roughly 100 cities in the United States—home to about 80 percent of the U.S. population—are out of compliance with air quality standards designed to protect public health and welfare. At the same time control costs now exceed $30 billion annually. Politicians are always debating next year's budget, Gregory J. McRae of Carnegie Mellon University has learned, and "in times of competing economic priority there is a terrible temptation to reduce the costs" of pollution study and control. An interesting problem for scientists, he believes, is to figure out how to lower both exposure to pollution and the costs of its control."

In his Frontiers symposium presentation, "Using Supercomputing and Visualization in Los Angeles Smog Simulation," McRae emphasized the growing need to bring knowledge of science to bear on the public policy domain. "People may try to pull the problems of pollution apart politically, but the underlying phenomena in fact are chemically coupled. We need to think about the chemical coupling at the time we are devising regulations," he maintained. Aware of pressing environmental problems, atmospheric scientists work with a sense of urgency to sort out and understand the intricacies of complex physical interactions. McRae, who has become a recognized authority on computer modeling of atmospheric phenomena, also "puts on his public policy hat" and spends an average of 3 days a month in Washington, D.C., translating and transmitting the science into its political context.

Explained McRae: "The role of the science is to try to understand what is going on in the atmosphere, and that involves three things: the chemical interactions that take place in the atmosphere, the emissions from mobile and stationary sources, and the meteorology that moves the materials around." The traditional role for atmospheric science has been, "given these inputs, to unravel the science that governs what happens, and to predict what will happen." Very often, however, the political process involves other nonscientific goals that inevitably color the way an atmospheric scientist works, often influencing his or her charge and data-gathering focus. "Atmospheric science is not practiced in isolation. The whole air-quality planning process involves science as well as social issues," according to McRae, who emphasized that scientists not only need to think about doing good science and engineering, but also need to venture a little beyond just the science and think about what that work and the results really mean to society. "With a bit of luck," he added in his presentation to the symposium's scientists, "we might perhaps convince you that that is an important thing to do."

McRae is a chemical engineer with what he described as "a fairly simple-minded view of the world, [the need for] what I call getting the sign right." To illustrate, he discussed chlorofluorocarbons (CFCs), engineered to be "extraordinarily good materials" and a very useful industrial product. But because of their particular chemical nature, CFCs turned out to have an unanticipated negative impact on stratospheric ozone. Millions of dollars were spent to create CFCs, but millions more will be spent to eliminate them and to mediate their adverse effect on the atmosphere. Thus there is a need, McRae emphasized, to go beyond concentrating on the physics and chemistry of how to make new materials and to think also about "the impact and pathways of chemicals after they leave the lab." McRae is also sensitive to the possibility that studies to elucidate phenomena in the atmosphere may not be balanced with a search for control strategies to deal with them. The urgency that attends the problems of pollution, he thinks, confers a responsibility on atmospheric scientists to consider issues like cost-effectiveness and adverse impacts whenever their results might be central to a political debate, which—in atmospheric science—is very often.

McRae has developed this awareness of the political arena as his model—begun in graduate school at Caltech—has evolved using a historical database of Los Angeles weather and pollutant data. That growing and coherent real-world test case permits him to evaluate what has now become a generic model, which he continues to develop, perfect, and apply from his present post at Carnegie Mellon University. The model's historical database provides useful comparative and baseline information for future simulations, of Los Angeles and of other urban applications. Because of the severity of pollution in Southern California, McRae and his model have been drawn deeply into the political decision-making process. Alan Lloyd, another session participant and chief scientist for the South Coast Air Quality Management District in California, emphasized that "it is critical to have available a wide group of scientists and institutions that regulators can turn to for unbiased and impartial information."

McRae credited the session's third participant, Arthur Winer, an atmospheric chemist at the University of California, Los Angeles, with "unraveling some of the very interesting things—the chemistry—going on at nighttime in the atmosphere." Winer brought to the symposium discussion his perspective as a longtime scientific advisor to California's regulators, and he also singled out as a challenge for atmospheric science the need "to arrest the declining numbers of really gifted young scientists" entering the field. He found it ironic that the surge in interest and academic programs in his field in the 1970s withered during the 1980s, just when dramatic environmental problems called for galvanizing society's resources.

HELPING REGULATORS DEFINE THE PROBLEM

Atmospheric scientists use the phrase *photochemical oxidant air pollution* to refer to a mixture of chemical compounds commonly known as smog. "The oxidants are not emitted directly, but rather are formed as products of chemical reactions in the atmosphere. It is this latter property that makes their control so difficult," explained McRae, because time of day and meteorological conditions directly affect the formation of pollution, as do the characteristics of the chemical compounds themselves. The precursors of smog are primarily nitrogen oxides, referred to by the symbol NO_x to indicate a sum of nitric oxide (NO) and nitrogen dioxide (NO_2), products of high-temperature fuel combustion, and reactive hydrocarbons resulting from solvent use, vegetative emissions, motor vehicles, and a broad spectrum of stationary sources.

Once the polluting sources have been identified, the next logical and traditional step taken by regulators has been to mediate the problem by reducing or redirecting the emitting source. But McRae pointed out one of the basic truths of this field that applies to most ecosystems: remedial actions may have unforeseen consequences. In the 1950s and 1960s, for example, regulators addressed the problem of pollution from coal-fired electrical power plants by mandating the construction of tall stacks—chimneys that soared hundreds of feet into the air to disperse the pollutants from the plants over a wide area. The "tall stacks solution" was counted a success because pollution near the power plants was dramatically reduced. Unfortunately, the sulfur-containing smoke from these plants, many of which were built in the American Midwest near high-sulfur coal deposits, traveled with the prevailing winds to New England and southeastern Canada. The material in the plumes then underwent chemical reactions and resulted in acid rain falling over large areas.

Thus, requiring tall stacks did not solve the pollution problem. Regulators were more successful, however, in responding to the problems of lead, which "is known to cause neurological malfunctioning, learning disabilities, and increased blood pressure. Almost all regulatory efforts in the past have been based on the simple notion that controlling the release of pollutant emissions will result in improved air quality. Regulatory activity designed to phase out the use of lead compounds as an antiknock additive in gasoline has been quite successful and has resulted in drastic reduction of ambient concentration levels" of lead in our air, said McRae. And so far the sign seems to be right: no significant or problematic unintended consequences have surfaced after the removal of lead from gasoline. Smog, however, is a larger, more complex problem.

REGULATION—ARE THE CORRECT QUESTIONS BEING POSED?

McRae and his colleagues have drawn some conclusions from their latest model that might influence regulators to redirect a current, major control strategy: the reduction of reactive organic gases. As with the tall stacks strategy, McRae wants to be sure "we aren't getting the sign wrong" and actually mandating a control strategy with unforeseen consequences that could be avoided. The environmental awareness that blossomed in the 1960s helped to establish the U.S. Environmental Protection Agency (EPA) in 1970. Atmospheric pollution was evident even before the dawn of the environmental movement: many cities were cloaked in a brown shroud of smog. Studies were sanctioned and conducted, and a series of regulatory guidelines developed to—again, logically—attack pollution at its source. With urban pollution, one major source appears to be the hydrocarbons that come from automobile engine combustion and from the use of solvents, together classified as reactive organic gases (ROGs). The EPA established a limit on these emission sources, and when McRae began modeling the atmosphere over Los Angeles in the early 1980s, one of the inevitable elements of his work was an evaluation of the effectiveness of lower emissions of ROGs. That evaluation, as it developed in the larger context of his model and in the work of other atmospheric scientists, surprised everyone.

"EPA's approach was basically right if you're looking very close to the source. But the role of hydrocarbons is essentially that of controlling the speed of the chemistry" that is leading to atmospheric pollution, said McRae. Other elements, primarily NO_x, turn out to be more instrumental in whether the pollution is actually generated. "If you put only organic controls on a city, it slows the reactions so that ozone forms downwind, outside the central city," McRae remarked. This generally means that regions and counties away from the center of a city—such as the Riverside and San Bernadino areas east of Los Angeles—are thus inundated with the smog. Because the formation of the pollutant was slower, "our national policy was leading to a situation where we may have reduced ozone in cities," but at the expense of suburban and rural areas downwind, so that "we were creating a problem on the regional scale." The sins of the generating region—as with the acid rain experience—were being visited upon nearby or faraway neighbors, depending upon the meteorology transporting them.

Running his model suggested to McRae an even more startling possibility. Regulating ROGs was not reducing ozone formation at all, because the role of NO_x had been misperceived and as a consequence, underestimated. Although, as Lloyd pointed out, "ozone levels in Los Angeles have been cut by about half over the last 15 years," the model indicated that in the rest of the country where NO_x was not being regulated, uncontrolled emissions of NO_x could confound the intended results of ROG reduction. "Despite massive control efforts," concluded McRae, "the average ozone levels in the eastern United States and in western cities such as Houston, Phoenix, and Denver have not been lowered. Part of the reason is that the chemistry of oxidant production is highly nonlinear. There is no simple relationship between the emissions of the major precursors and the resulting ozone."

Lloyd noted that the EPA is now turning to control of NO_x as well as ROGs, an approach taken early on by regulators in California, and he emphasized as well the amount of work still to be done by scientists in addressing problems of air pollution. Especially important, he maintained, is the need to obtain adequate amounts of high-quality experimental data, the lack of which will compromise even the best efforts to model and understand complex physical and chemical phenomena basic to environmental change.

CHEMISTRY IN THE ATMOSPHERE

Ozone as a Protective Shield

Ozone formation in the lower atmosphere may be one of the more popularly misunderstood of scientific phenomena, but its importance to the Earth's fate cannot be overestimated. Part of the popular confusion involves a crucial distinction that must be kept in mind: the environmental impact of ozone depends on the altitude at which one considers it. In *One Earth, One Future* (Silver and DeFries, 1990), authors Cheryl Simon Silver and Ruth S. DeFries explain (pp. 104–109) that ozone tends to accumulate in a layer in the Earth's stratosphere, with the maximum concentrations occurring at a height of roughly 25 to 35 kilometers above the Earth's surface. Although it is only a small component of the Earth's stratosphere, ozone nonetheless performs a vital function: the layer where it accumulates filters the Sun's rays and blocks out much of the radiation in the ultraviolet-B wavelength range, thereby contributing to an environment on the planet's surface where life could and did evolve. Even though its density averages only 3 parts per million (ppm) throughout the atmosphere and at its stratospheric peak only about 9 ppm, at that level it screens out more than 99 percent of the potentially damaging radiation.

Oxygen as a molecule is composed of two atoms (O_2); ozone (O_3) is a highly reactive molecule composed of three oxygen atoms. The atmosphere is rich with O_2 molecules, which in the upper stratosphere are exposed to large amounts of the Sun's radiant energy that is relatively unaffected by its journey through space. Photolysis is the process whereby this energy is absorbed by a reactive chemical species that changes as a result. Once split into two atoms by photolysis, oxygen provides the raw materials, because these liberated atoms can then join with other O_2 molecules to form ozone. The so-called ozone layer of the stratosphere with the heaviest concentrations develops at the altitudes it does because of a natural balance: higher up, above the stratosphere, the density of oxygen is such that liberated oxygen atoms encounter relatively few oxygen molecules to join to; lower down in the troposphere, insufficient light energy arrives due to the shielding effect of the ozone that has been created above.

When McRae referred to engineers failing to "get the sign right" with CFCs, he was referring to damage done to the ozone layer by this man-made chemical creation (Box 8.1). CFCs are classified as synthetic halocarbons, and by photolysis can be broken down to release fluorine and chlorine atoms. When liberated, a single chlorine atom can trigger a catalytic reaction and destroy thousands of ozone molecules, thus contributing to a breakdown of the Earth's protective ozone shield.

Ozone as a Pollutant

Ozone in the upper stratosphere is created largely by natural processes, and its presence serves a vital function. Lower down in the troposphere, however, ozone is classified as a pollutant, with adverse health effects. "In the Earth's atmosphere, the primary tendency of the chemistry is to oxidize any molecules emitted into the air," said McRae, who pinpointed ozone as one of the two primary species that mediate the process. Published work by McRae and Armistead G. Russell discusses some of the chemistry basic to models of air pollution and explains some of the primary phenomena (Box 8.2).

BOX 8.1 CHLOROFLUOROCARBONS

CFC molecules consist of either a single carbon atom or a pair of carbon atoms bonded to a fluorine and a chlorine atom, and have an atomic structure that confers on them a collection of beneficial properties. They are light in weight, neither flammable nor toxic, and are largely impervious to degradation by microorganisms or to reactions with other chemicals. Their applications are many: refrigerants in air conditioners and refrigerators, propellants for aerosol cans, foam blowing agents, and solvents for cleaning electronic components and other industrial products, to name only the most significant. But these same chemical qualities make them problematic, McRae explained. Since they do not react or degrade and are lighter in weight than nitrogen or oxygen, once they escape from their earthbound containers or have served their utilitarian function, they endure and migrate very slowly over a number of years or decades into the stratosphere.

It is upon their eventual arrival in the stratosphere that CFCs turn deadly—in McRae's terminology, reverse their positive sign. Again, photolysis occurs when radiant light energy reacts with valence electrons to dissociate atoms and break down chemical bonds.

BOX 8.2   OZONE PRODUCTION IN THE TROPOSPHERE

Ozone production in the troposphere comes from the photolysis of nitrogen dioxide (NO_2),

$$NO_2 + light \rightarrow NO + O(^3P),\qquad(1)$$

which generates nitric oxide (NO) and a species of oxygen atom, $O(^3P)$, that is highly reactive, referred to as the triplet state. In the presence of a spectator molecule like nitrogen (N_2) or oxygen (O_2)— designated M in this reaction—the triplet oxygen reacts with molecular oxygen to form ozone (O_3):

$$O(^3P) + O_2 \overset{M}{\rightarrow} O_3.\qquad(2)$$

In turn, the ozone can react with the nitric oxide produced in the first step

$$O_3 + NO \rightarrow NO_2 + O_2\qquad(3)$$

to produce nitrogen dioxide and molecular oxygen, thus circling back to the original atmospheric components. In this cycle of reactions, no net ozone is produced.

Under typical atmospheric conditions, McRae said, the cycle is very rapid, so that the three key species are approximately in equilibrium:

$$[O_3] = \kappa\frac{[NO_2]}{[NO]}\qquad(4)$$

"Since most nitrogen oxide emissions are in the form of nitric oxide (NO), these relationships suggest that ozone levels should be quite low, on the order of 0.02 ppm," according to McRae and Russell. They noted that the levels that are actually observed can be an order of magnitude higher than this. Los Angeles has peak ozone levels above 0.30 ppm, and Mexico City's levels are often higher. Thus, "one of the key questions in atmospheric photochemistry is to understand how ozone can accumulate to these high levels in urban and rural regions. Part of the answer," the authors said, derives from the other primary atmospheric species involved in oxidation, the hydroxyl radical.

The hydroxyl radical is chemically adept at oxidizing organic species present in the urban atmosphere and as a consequence is largely responsible for elevated ozone levels in and around cities. Hydroxyl radicals can be produced in several ways. When an ozone molecule absorbs a relatively high-energy photon, it dissociates to form molecular oxygen and a relatively long-lived, electronically excited atom known as the singlet D state (^1D):

continued on next page

BOX 8.2—*continued*

$$O_3 + \text{light} \rightarrow O(^1D) + O_2. \qquad (5)$$

This species can react with water vapor to form two hydroxyl radicals:

$$O(^1D) + H_2O \rightarrow 2OH. \qquad (6)$$

"A second source of OH radicals," they said, "is from oxygenated organics. Formaldehyde (HCOH) provides the simplest example," using photolysis to produce hydroperoxyl radicals (HO_2):

$$HCOH + \text{light} \rightarrow H + HCO, \qquad (7)$$

$$H + O_2 \xrightarrow{M} HO_2, \qquad (8)$$

and

$$HCO + O_2 \xrightarrow{M} HO_2 + CO. \qquad (9)$$

These HO_2 radicals themselves can interact to produce hydrogen peroxide (H_2O_2), which in turn can also undergo photolysis to produce OH radicals":

$$HO_2 + HO_2 \rightarrow H_2O_2 + O_2 \qquad (10)$$

and

$$H_2O_2 + \text{light} \rightarrow 2OH. \qquad (11)$$

Thus the generation of hydroxyl radicals is understood, but less apparent is why they should generate excess ozone, since the basic equilibrium equation (4) is a function of the ratio of nitrogen dioxide to nitric oxide. The answer is in the propensity of the hydroperoxyl radicals to interact with and oxidize nitric oxide:

$$HO_2 + NO \rightarrow OH + NO_2. \qquad (12)$$

The products of this reaction are important. Not only are more hydroxyl radicals produced, but newly formed nitrogen dioxide also pushes the numerator of the ozone equilibrium equation up, and along with it the concentration of ozone. Without these excess concentrations, it normally requires an ozone molecule to oxidize nitric oxide into nitrogen dioxide, but equation (12) describes a reaction that, unlike equation (3), "provides a way to oxidize NO without consuming an ozone molecule, which in turn implies that the equilibrium in (4) is shifted to higher concentration levels," according to McRae and Russell.

SOURCE: McRae and Russell (1990).

To summarize the chemistry: certain photolytically driven reactions (such as the interaction of hydroxyl radicals with organics) tend to oxidize NO to NO_2 and/or to produce more hydroxyl radicals. Such chemical reactions can contribute to a cycle or chain reaction that intensifies the concentration of ozone, and hence increases smog. Knowing how to identify and apply the reactions is the important framework, but obtaining accurate measurements of these substances— which can exist in a transient condition, swirling around the atmosphere, in concentrations as small as a few parts per trillion—presents a formidable challenge to atmospheric chemists.

GETTING THE NUMBERS RIGHT

Winer necessarily brings a historical perspective to any consideration of atmospheric science. For 20 years he has been a prominent participant in a number of important developments in the field, watching the science of measurement, the use of computers, and the role of the scientist in the public forum all evolve. In his early work, Winer was concerned "that too much attention was being given to six so-called criteria pollutants under the Clean Air Act" in the mid- to late-1970s and that as a consequence not much was known about a whole spectrum of other species such as formaldehyde (HCHO), nitrous acid (HONO), nitric acid (HNO_3), the nitrate radical (NO_3), dinitrogen pentoxide (N_2O_5), and formic acid (HCOOH). "These species either had never been measured in the atmosphere, or had, at best, been detected by wet chemical methods of notorious unreliability," said Winer, who added that, "not surprisingly, almost nothing was known about the possible health impacts of these compounds." Then at the Statewide Air Pollution Research Center (SAPRC), a research institute of the University of California based on the Riverside campus, Winer and a SAPRC team led by James N. Pitts, Jr., began a series of experiments to "try to understand what's really *in* a smog cloud," using Fourier transform infrared (FT-IR) spectroscopy for the first time over long optical paths of a kilometer or more. The parts-per-billion detection sensitivities afforded by such pathlengths led to the first spectroscopic detection of HCHO and HNO_3 in polluted air.

This success spurred Winer and his colleagues to collaborate with German physicists Uli Platt and Dieter Perner, who had devised a novel spectroscopic instrument operating in the UV-visible region. By exploiting long optical paths—up to 17 kilometers—and a rapid scanning computer-based capability in which a signal-averaging algorithm was applied to thousands of superimposable spectral records, Winer, Platt, and their collaborators were able to achieve parts-per-trillion detection limits and hence the first observation of the nitrate radical and nitrous acid in the polluted troposphere (Platt et al., 1980a, b). Thus the SAPRC-German teams established the methodological basis for later, more sophisticated models predicated on accurate measurements of atmospheric components. They worked on what Winer called "an historical antecedent" to McRae's powerful model, based on the premise that the more definitive the measurable data used as input, the more likely the model will be to mimic real-world processes.

NIGHTTIME CHEMISTRY

The long optical-path measurements by the SAPRC-German collaborators opened inquiry into an area that had been previously overlooked, nighttime chemistry. Sunlight drives most of the chemical reactions that atmospheric scientists model, and thus conventional wisdom held that nighttime—with its absence of hydroxyl radicals— might offer a respite from the creation of secondary pollutants. General pollution measurements taken after dark seemed to confirm that nighttime chemistry "was of minor importance," said McRae, who added that it "was even disregarded in some models." NO_3, for example, is a trace nitrogen-containing radical with such small predicted concentrations (in the parts-per-trillion range) in the troposphere that scientists largely ignored its possible significance. But Platt, Winer, and Pitts and their co-workers focused their special spectrograph (Figure 8.1) on just such overlooked yet potentially significant species. They were the "first to discover the nitrate radical in the polluted troposphere," and then went on to detect it in the cleaner troposphere as well, said McRae. "We have now shown systematically in cleaner atmospheres that after sunset the nitrate radical concentration grows into more or less steady-state levels," said Winer, "which persist throughout the night and disappear at sunrise due to photolysis" (Platt et al., 1984).

This discovery, notwithstanding the small amount of the nitrate radical they found, could hardly have been more confounding and significant to the atmospheric sciences community. Suddenly, the formation of nitric acid in the troposphere at night could be explained. Previously, it was believed that nitric acid could be produced only by the reaction of the hydroxyl radical with NO_2 during daylight hours. With the discovery of the nitrate radical concentration arising after dark, the leftover pollutants O_3 and NO_2 were provided a pathway leading to the nighttime production of nitric acid.

Winer, as a Californian, immediately recognized the implications of this discovery, since "nitric acid is, in fact, two-thirds of our acid

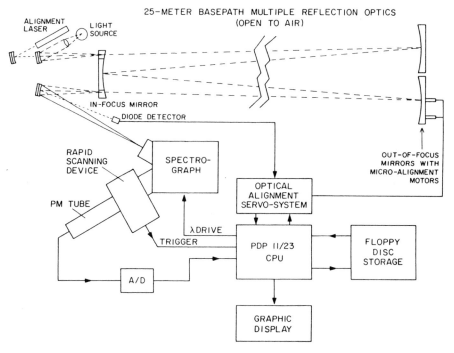

FIGURE 8.1 Schematic diagram of differential optical spectroscopy absorption systems used to measure atmospheric concentrations of the nitrate radical. (Courtesy of A.M. Winer.)

rain problem in the western United States. We don't burn many high-sulfur fuels, and consequently don't have much of the sulfur dioxide problem" that is so prominent in the East. The nitrate radical held even more significance, however, once its nighttime presence was established. Scientists had long puzzled over the absence of accumulated nighttime concentrations of compounds emitted from vegetation in large amounts. Isoprene and the monoterpenes were shown to react rapidly and efficiently with the nitrate radical, which effectively becomes a nighttime sink to mediate their buildup in the atmosphere (Winer et al., 1984).

The final discovery was even more significant, said Winer, solving "a longstanding problem in marine science." Dimethyl sulfide (DMS) is produced in large quantities from the world's oceans but does not accumulate during the day because of the presence and interaction of the hydroxyl radical. Again, since the OH radical disappears at night, atmospheric chemists were unable to explain why

DMS concentrations did not swing higher after dark. The nitrate radical turns out to be extremely efficient at titrating DMS, said Winer, and thus suppresses its buildup over the oceans at night (Winer et al., 1984).

The spectroscopic measurements made by Winer and his colleagues explained one final piece of the nighttime puzzle, which is suggestive of a larger pattern of diurnal pollution cycling that may be fairly independent of the meteorology. The ubiquitous hydroxyl radical, he said, "is the single most important reactive intermediate in the atmosphere. It catalyzes and drives smog formation." Winer and his colleagues confirmed the accumulation of substantial amounts of nitrous acid (HONO) at night in the Los Angeles basin. Since HONO is a photolytic species that readily breaks apart into the hydroxyl radical and nitric oxide during the day, Winer continued, "the accumulated nitrous acid that photolyzes at sunrise leads to a pulse of OH radicals that represents a kicking off of the next day's smog." In the fall and winter seasons in California when ozone and formaldehyde (which also photolyze to produce OH radicals) are reduced, this OH radical kickoff by nitrous acid can become a dominant source of OH radicals.

McRae clarified for the symposium three important implications from discoveries by Winer, Pitts, and others that have helped to redefine the urban pollution problem. First, nitric acid can be created in a way that requires neither sunlight nor OH radicals; second, he said, "previous model simulations that did not include this reaction would underpredict the importance of controlling NO_x emissions; and finally, peroxyacetyl nitrate (PAN) can lead to long-distance transport of reactive nitrogen," a phenomenon that had been known about previously but was further clarified here.

And they also demonstrate the utility of computer models, which, in this case, revealed a crucial part of the picture that scientists had overlooked. Compared to the immediate troposphere where pollutants often accumulate in high and easily detectible concentrations, the upper atmosphere where these reactions occur was harder to observe. Thus the trace concentrations of the nitrogen-containing NO_3 radical, to most scientists, had seemed less significant to their models than had the more apparent phenomena. Computer models can be constructed to exclude such judgmental criteria, however. Therefore serendipities can emerge such as this detection of a chemically rapid path by which NO_3 can produce nitric acid and PAN without the sunlight's driving energy, especially in the upper portion of the polluted layer. There is something particularly compelling about such discoveries, which seem bred of the inherent complexities of a dynamical system like the atmosphere when explored with the comparably inexhaustible power of the computer.

COMPUTATIONAL EXPERIMENTATION

McRae appreciates that this use of the computer represents a new stage in the history of the scientific method: "This is where high-performance computing has played a very critical role, because we can now begin to think in terms of computational experiments. Traditionally, in the past, scientists had a theoretical formulation and then went into the lab to verify it." But meteorology has always been difficult to fit into that tradition because one cannot simply design and conduct experiments in the laboratory of a real city. With advanced models like the one McRae has spent the better part of a decade developing and refining, however, atmospheric science is beginning to utilize the experimental method more as some of the hard sciences do.

An atmospheric scientist may bring many theoretical considerations to bear in creating a model, but it receives its pedigree only by successfully matching nature. That is, given as inputs for its calculations the measurements and descriptive weather data that were actually observed at a given historical period, the model will predict a pattern of change. This simulation of an evolving weather and pollution scenario is then compared to the conditions that transpired and were measured, and the model's credibility can thereby be established. "Before a model can be used for control strategy calculations, it is critically important to demonstrate that it has the capability to accurately reproduce historical events," emphasized McRae. Once confirmed as a descriptive model, it has license as a prescriptive model. "You can with more confidence use the model in a predictive mode, the traditional role of science," said McRae. Lloyd cautioned, however, against overstating the reach of these newer methods: "Much more work is still needed to understand, for example, the meteorology and turbulence of the atmosphere. We can still only approximate."

MORE THAN A TOOL, THE COMPUTER
PROVIDES A FRAMEWORK

In the political forum McRae makes use of some very illustrative video presentations of his work—which he showed to the symposium—that convey the dynamics of the atmosphere with a power and coherence undreamed of before the development of the sophisticated modeling and visualization techniques that he and others now rely on. Atmospheric modeling on computers has proven to be a key to unlock possibilities that previous generations of scientists could not imagine. McRae's use of the computer provides a vivid illustration of how a working scientist can master this revolutionary tool to the point where his power to theorize and to predict virtually redefines the scope of problems he can tackle. Lloyd believes that McRae's greatest contributions to the field have been made as a modeler.

The session's organizer, Larry Smarr, who directs the National Center for Supercomputing Applications (NCSA) at the University of Illinois, reminded the symposium audience that McRae was the first winner, in 1989, of the Frontiers of Computational Science Award. Writing about McRae and other scientists breaking new ground with computers, Robert Pool (1984) anticipated a theme that was to be visited during the Computation session at the Frontiers symposium: "Not only can computer experiments substitute for many of the studies normally done in a lab, but they also are allowing scientists to gather data and test hypotheses in ways that were closed to them before. And the field is still in its infancy—as computers become more powerful and as more scientists become aware of their potential, computer experiments are likely to change the way research is done. The promise of computational science has led some researchers to suggest the field will eventually grow into a third domain of science, coequal with the traditional domains of theory and experimentation" (p. 1438).

McRae came to the California Institute of Technology from Melbourne, Australia, in the mid-1970s for graduate work on control theory, but experienced a sea change that altered his career path. "Once you fly into Los Angeles, you realize there's a very severe problem there," he recalled. He began working on a computational model of the chemistry and physics of air pollution, which in the dissertation stage brought him into contact with mechanical engineer Armistead G. Russell. (The two are now colleagues at Carnegie Mellon, where they make good use of the Pittsburgh Supercomputing Center's CRAY-YMP 8/832.) The model contains the formulas and equations necessary to capture the relevant variables and relationships based on what is understood about the atmospheric physics and chemistry. McRae conceived a set of hypotheses in the form of a strategy to reduce pollution. He set the emissions levels of the target pollutants at variable levels and then evolved their chemical interaction through time under the influence of the other major set of variables, the meteorology (wind speed and direction, cloud cover, temperature, and so on).

While he was learning more about modeling the atmosphere, he was also collecting more and more data from Los Angeles air monitoring studies. Thus, moving to Pittsburgh did not deflect the focus of his model, but did expand it into the supercomputing environment, where the Cray he uses harnesses eight processors that can perform up to 2.7 billion calculations per second at its peak capacity. With this engine to drive it, McRae's model makes far fewer concessions (than are made in some other models) to the inherent limitations and extrapolations necessary to deal with nonlinear mathematics. The Navier-Stokes equations are a complex set of such equations that have been developed to model the weather. McRae created a grid map of the air over the entire Los Angeles basin that marked out 12,000 so-called mesh points, where all of the relevant interactions and phenomena would be individually computed. (Smarr has noted that only a few historical data points exist to provide pollution input numbers to drive the predictive model.) Fifty chemical species and their interactions are modeled at each of these 12,000 points, and the equations are then coupled. The resulting model staggers the imagination, in fact would never have even been imagined before the advent of supercomputers: 500,000 coupled nonlinear equations, which McRae then manipulated into 200 different simulations, each for a different set of environmental conditions. The computer can run through one of these full 2-day simulations in about 40 minutes.

With this vast sea of data under control, McRae worked with graphics specialists at NCSA to develop a computer-generated video presentation that uses scientific visualization techniques to illuminate the results of the model. Simulations of Los Angeles, over several days under specific environmental conditions, can be compressed into a few moments with modern visualization tools (Figure 8.2). But now that the model has earned its stripes as a predictor, it is the scientists who are defining the levels of various hypothetical pollutant emissions and the weather influencing them. The model then produces a vision of what such pollution and weather in Los Angeles would look like that has been captured in a video, produced by the Pittsburgh Supercomputing Center, called "Visualizing Los Angeles Air Quality." A second video called "Smog" was developed and produced by the National Center for Supercomputing Applications "to try to understand the physics and the chemistry," where the heart of McRae's work lies. "We can work backwards from a condition, unraveling the meteorological transport. If this is the worst air quality experienced, what are the sources that might be responsible for it?" he asks. Armed with these tools, then, McRae, his colleagues,

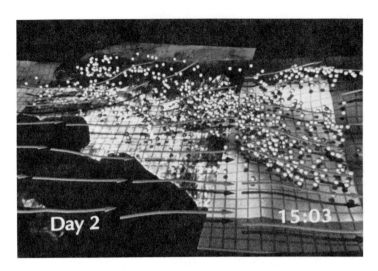

FIGURE 8.2 Computer-generated simulation of the air over Los Angeles. (Courtesy of L.L. Smarr.)

and other atmospheric scientists venture forth from their laboratories into the pragmatic, political world of pollution control.

REFRAMING THE LAWS OF POLLUTION CONTROL

McRae's long history of involvement with colleagues in the field and regulators in Southern California, along with the demonstrable cogency of his methods and his model, have given him a role in recent regulatory reform. Said Lloyd, "His work gave technical credence to strong regulations that were needed to reduce emissions." The South Coast Air Quality Management District, said Lloyd, is well served by the views and resources provided by individuals like Winer and McRae, although Lloyd's premise is that "the scientists and the scientific community have to be more proactive." Thus, when the model McRae has spent nearly a decade developing and refining was finally put to the test, and its results suggested many fairly dramatic implications, the battle was just being joined. Industry and other interested parties suggested, supported, and then promoted alternative studies that turned out to contradict McRae's work, and a fairly heated political and media debate ensued, fueled by major lobbying interests that had a lot at stake.

Ultimately, the decisions to regulate ROGs and NO_x that California's legislators and regulators had previously made were ratified by the superiority and demonstrable conclusions of McRae's and Russell's work, which further buttressed legislative and policy plans for the state's environmental strategy. McRae pointed out one of the essential ingredients in this vindication of their model: their determination to certify the work as solid science. "[The industry modelers] simply claimed their conclusions without exposing the details of the calculations to scientific peer review," McRae was quoted as saying (Schneider, 1990, p. 44). McRae and his colleagues, however, invited the test of scientific scrutiny by publishing their results in an established journal (Milford et al., 1989). Although purely scientific, that paper indicates the nature of pollution politics and reads like a manifesto for change, based on the strong demonstrative evidence developed by the authors from the model of Los Angeles. The following paraphrase of that paper matches the tone and summarizes the contents of the published version:

The Problem. More than half of the people living in the United States reside in areas where ozone concentrations exceed the National Ambient Air Quality Standard of 0.12 ppm mandated by the U.S. government, and many cities, like Los Angeles, also fail to comply with the standards governing particulate matter (PM_{10}) pollution. In the face of this shortcoming, science suggests that the standards should be even tighter, and many agencies already feel that they have exhausted the technical, political, and economically feasible control measures currently available to address the problem. Moreover, there is a growing awareness that the scope of the problem is not fully captured by the historical convention of concentrating on cities, since recent evidence and modeling suggest downwind problems are sometimes more severe than in the areas whence the pollution was generated.

The Traditional Response. Given a mandated, numerical air quality goal, strategies to reduce ROG emissions to a level that leads to air that meets that standard have been evaluated for cost-effectiveness. The Empirical Kinetics Modeling Approach employs a simplified, single-cell trajectory formulation to model the impact of a number of possible ROG and NO_x emission level combinations on the resultant ozone level, and then to derive a new ROG reduction goal. Lowering NO_x has not been mandated (except in some jurisdictions such as California), since observations suggest that sometimes lower NO_x can paradoxically lead to higher ozone. The model fails to incorporate the significance of other pollutants, such as gas-phase pollutants other than ozone, and particulate matter components less than 10 micrometers in diameter. Further, it is not clear whether peak ozone concentrations observed within the urban area are the best measures of air quality for use in designing control strategies.

A New Modeling Approach. The use of new Cray Research super-computers permits simulations predicated on a different approach to modeling that were formerly computationally infeasible. Current photochemical airshed models accurately describe the physical and chemical processes responsible for chemical transformation, transport, and fate. Advances in computational technology now permit the exploration of more control scenarios at a greater number of sites at smaller time intervals. Considering the entire Los Angeles basin as a rectangular grid 200 by 100 kilometers, 45 different combinations of control levels (most of them limited to 80 percent or less of an improvement over current standards, those more likely to be implemented over the next decade) were simulated under expected atmospheric conditions for thousands of adjacent but discrete locations within the shed over a 2-day period, each simulation requiring 40 minutes of computer time. The choice of issues and methods was designed to widen the applicability of the model to other urban locations.

The model must first demonstrate its reliability by reproducing air quality outcomes and readings for O_3, NO_2, HNO_3, NH_4NO_3, and PAN that were historically experienced. August 30 and 31, 1982, were chosen for this purpose for a number of meteorological reasons. When compared to the data collected for those days, the model's performance was in excellent agreement for ozone and PAN, and for nitrogen dioxide was better than with previous models. Also, the gas-phase and particulate predictions were in excellent agreement with observations. This base-case performance established the model's reliability, and simulations of differing control strategies were run.

The Results and Implications for Control Strategies. A number of surprising spatial patterns were observed for the first time, as this modeling technique revealed interactions and time-sensitive patterns for the targeted substances that had not been captured by previous models. In particular, the model demonstrated that changing relative emissions standards very often had the impact not of reducing peak ozone concentrations, but of merely moving them to a different region within the airshed. Another strongly supported observation involves the effect of reductions of emissions on concentrations of other pollutants, such as PAN and inorganic nitrate. The modeling seems to suggest several important premises for regulators trying to balance the cost and effectiveness of control strategies:

• Trade-offs will be inevitable, because various choices among control strategies will have differential impacts on different receptor sites throughout the region. Also, the ratio of ROG to NO_x reductions will affect pollutants other than ozone, in differing degrees, and for different receptor sites.

• NO_x reductions may actually increase ozone in regions of the airshed with high emissions. But as receptor sites further downwind are considered, the impact of reducing NO_x throughout the basin tends to reduce ozone. The crossover point seems to be 3 to 4 hours of downwind travel time. Associated ROG controls are most effective in those areas where high NO_x levels are maintained and radical concentrations suppressed through midday.

• PAN concentrations seem to follow the same general pattern as ozone. Thus a combination of ROG and NO_x controls—such as have been in effect for about a decade in California—is advisable.

• Inorganic nitrate levels follow base-line levels of NO_x and OH. Thus reducing NO_x reduces inorganic nitrate pollution.

• The foregoing points suggest that a strategy solely reliant on ROG emission reductions may be inadequate, as the EPA now recognizes, and under certain conditions could never succeed in attaining the required ozone standard, without correlative substantial reductions in NO_x. Moreover, only by a strategy that combines both ROG and NO_x controls can simultaneous reductions be expected in ozone, PAN, and inorganic nitrates.

• Any conclusions drawn about the effects of pollution based only on peak receptor sites will distort the picture that emerges from more revealing models. These suggest that only by considering numerous receptor sites throughout the region and over time, and evaluating at least PAN and inorganic nitrates in addition to ozone, can a strategy be developed that embraces the known major trade-offs.

SCIENCE IN THE PUBLIC INTEREST

Thus it can be seen from the work authored by Milford, Russell, and McRae that atmospheric scientists very often assume a strongly proactive stance about their conclusions. As McRae affirmed at the outset of his presentation at the Frontiers symposium, convincing people about the right thing to do in response to the problems he is analyzing is part of his charge. He and Winer and Lloyd each in his own way revealed a consciousness informed by science, but fueled with a passion for implemented solutions. And as McRae's career and his model attest, modern supercomputing provides a very useful

and powerful tool that has helped to revolutionize the study of meteorology and to create a more comprehensive vision of atmospheric
science.

BIBLIOGRAPHY

McRae, Gregory J., and Armistead G. Russell. 1990. Smog, supercomputers,
and society. Computers in Physics (May/June):227-232.
McRae, G.J., J.B. Milford, and J.B. Slompak. 1988. Changing roles for supercomputing in chemical engineering. International Journal of Supercomputing Applications 2:16-40.
Milford, Jana B., Armistead G. Russell, and Gregory J. McRae. 1989. A new
approach to photochemical pollution control: Implications of spatial
patterns in pollutant responses to reductions in nitrogen oxides and reactive organic gas emissions. Environmental Science and Technology
23(10):1290-1301.
Platt, U., D. Perner, A.M. Winer, G.W. Harris, and J.N. Pitts, Jr. 1980a. Detection of NO_3 in the polluted troposphere by differential optical absorption. Geophysical Research Letters 7:89-90.
Platt, U., D. Perner, G.W. Harris, A.M. Winer, and J.N. Pitts, Jr. 1980b. Observation of nitrous acid in an urban atmosphere by differential optical
absorption. Nature 285:312-314.
Platt, U., A.M. Winer, H.W. Biermann, R. Atkinson, and J.N. Pitts, Jr. 1984.
Measurements of nitrate radical concentrations in continental air. Envi-
ronmental Science and Technology 18:365-369.
Pool, Robert. 1989. Is it real, or is it Cray? Science 244:1438-1440.
Schneider, Mike. 1990. Visualizing clean air in Los Angeles. Supercomput-
ing Review 3(March):44-48.
Silver, Cheryl Simon, and Ruth S. Defries. 1990. One Earth, One Future.
National Academy Press, Washington, D.C.
U.S. Environmental Protection Agency. 1988. Review of the National Ambi-
ent Air Quality Standards for Ozone. Staff Paper. Office of Air Quality
Planning and Standards. EPA, Research Triangle Park, N.C.
Warneck, P. 1988. Chemistry of the Natural Atmosphere. Academic Press,
San Diego, Calif.
Winer, A.M., R. Atkinson, and J.N. Pitts, Jr. 1984. Gaseous nitrate radical:
Possible nighttime atmospheric sink for biogenic organic compounds.
Science 224:156-159.

RECOMMENDED READING

Finlayson-Pitts, Barbara, and James Pitts. 1986. Atmospheric Chemistry—
Fundamentals and Experimental Techniques. John Wiley and Sons, New
York.
Seinfeld, John. 1986. Atmospheric Chemistry and Physics of Air Pollution.
John Wiley and Sons, New York.

9

Computational Neuroscience: A Window to Understanding How the Brain Works

"The brain computes!" declared Christof Koch, who explained at the Frontiers of Science symposium how a comparatively new field, computational neuroscience, has crystallized an increasingly coherent way of examining the brain. The term *neural network* embraces more than computational models. It now may be used to refer to any of a number of phenomena: a functional group of neurons in one's skull, an assembly of nerve tissue dissected in a laboratory, a rough schematic intended to show how certain wiring might provide a gizmo that could accomplish certain feats brains—but not conventional serial computers—are good at, or a chip intended to replicate features of any or all of these using analog or digital very-large-scale integrated circuit technology. Neural networks are more than a collection of pieces: they actually embody a paradigm for looking at the brain and the mind.

Only in the last decade has the power of neural net models been generally acknowledged even among neurophysiologists. The paradigm that now governs the burgeoning field of computational neuroscience has strong roots among both theoreticians and experimentalists, Koch pointed out. With the proviso that neural networks must be designed with some fidelity to neurobiological constraints, and the caveat that their structure is analogous to the brain's in only the very broadest sense, more and more traditional neuroscientists are finding their own research questions enhanced and provoked by neural net modeling. Their principal motivation, continued Koch, "is the realization that while biophysical, anatomical, and physiological data are necessary to understand the brain, they are, unfortunately, not sufficient."

What distinguishes the collection of models and systems called neural networks from "the enchanted loom" of neurons in the human brain? (Pioneer neuroscientist Sir Charles Sherrington coined this elegant metaphor a century ago.) That pivotal question will not be answered with one global discovery, but rather by the steady accumulation of experimental revelations and the theoretical insights they suggest.

Koch exemplifies the computational neuroscientist. Trained as a physicist as opposed to an experimental neurobiologist, he might be found at his computer keyboard creating code, in the laboratory prodding analog very-large-scale integrated (VLSI) chips mimicking part of the nerve tissue, or sitting at his desk devising schema and diagrams to explain how the brain might work. He addressed the Frontiers symposium on the topic "Visual Motion: From Computational Analysis to Neural Networks and Perception," and described to his assembled colleagues some of the "theories and experiments I believe crucial for understanding how information is processed within the nervous system." His enthusiasm was manifest and his speculations about how the brain might work provocative: "What is most exciting about this field is that it is highly interdisciplinary, involving areas as diverse as mathematics, physics, computer science, biophysics, neurophysiology, psychophysics, and psychology. The Holy Grail is to understand how we perceive and act in this world—in other words, to try to understand our brains and our minds."

Among those who have taken up the quest are the scientists who gathered for the Frontiers symposium session on neural networks, most of whom share a background of exploring the brain by looking at the visual system. Terrence Sejnowski, an investigator with the Howard Hughes Medical Institute at the Salk Institute and University of California, San Diego, has worked on several pioneering neural networks and has also explored many of the complexities of human vision. Shimon Ullman at the Massachusetts Institute of Technology concentrates on deciphering the computations used by the visual system to solve the problems of vision. He wrote an early treatise on the subject over a decade ago and worked with one of the pioneers in the field, David Marr. He uses computers in the search but stated his firm belief that such techniques need to "take into account the known psychological and biological data." Anatomists, physiologists, and other neuroscientists have developed a broad body of knowledge about how the brain is wired together with its interconnected neurons and how those neurons communicate the brain's fundamental currency of electricity. Regardless of their apparent success, models of how we think that are not consistent with this knowledge engender skepticism among neuroscientists.

Fidelity to biology has long been a flashpoint in the debate over the usefulness of neural nets. Philosophers, some psychologists, and many in the artificial intelligence community tend to devise and to favor "top-down" theories of the mind, whereas working neuroscientists who experiment with brain tissue approach the question of how the brain works from the "bottom up." The limited nature of top-down models, and the success of neuroscientists in teasing out important insights by direct experiments on real brain tissue, have swung the balance over the last three decades, such that most modelers now pay much more than lip service to what they call "the biological constraints." Also at the Frontiers symposium was William Newsome. Koch described him as an experimental neurophysiologist who embodies this latter approach and whose "recent work on the brains of monkeys has suggested some fascinating new links between the way nerve cells behave and the perception they generate."

Increasingly, modelers are trying to mimic what is understood about the brain's structure. Said James Bower, a session participant who is an authority on olfaction and an experimentalist and modeler also working at the California Institute of Technology, "The brain is such an exceedingly complicated structure that the only way we are really going to be able to understand it is to let the structure itself tell us what is going on." The session's final member, Paul Adams, an investigator with the Howard Hughes Medical Institute at SUNY, Stony Brook, is another neuroscientist in the bottom-up vanguard looking for messages from the brain itself, trying to decipher exactly how individual nerve cells generate the electrical signals that have been called the universal currency of the brain. On occasion, Adams collaborates with Koch to make detailed computer models of cell electrical behavior.

Together the session's participants embodied a spectrum of today's neuroscientists—Koch, Sejnowski, and Ullman working as theoreticians who are all vitally concerned with the insights from neurobiology provided by their experimentalist colleagues, such as Bower and Adams. From these diverse viewpoints, they touched on many of the elements of neural nets—primarily using studies of vision as a context—and provided an overview of the entire subject: its genesis as spurred by the promise of the computer, how traditional neuroscientists at first resisted and have slowly become more open to the promise of neural networks, how the delicate and at times hotly debated interplay of neuroscience and computer modeling works, and where the neural net revolution seems headed.

COMPUTATION AND THE STUDY OF THE NERVOUS SYSTEM

Over the last 40 years, the question has often been asked: Is the brain a computer? Koch and his colleagues in one sense say no, lest their interlocutors assume that by computer is meant a serial digital machine based on a von Neumann architecture, which clearly the brain is not. In a different sense they say yes, however, because most tasks the brain performs meet virtually any definition of what a computer must do, including the famous Turing test. English mathematician Alan Turing, one of the pioneers in the early days of computation, developed an analysis of what idealized computers were doing, generically. Turing's concept employed an endless piece of tape containing symbols that its creators could program to either erase or print. What came to be called the Turing machine, wrote physicist Heinz Pagels, reduces logic "to its completely formal kernel. . . . If something can be calculated at all, it can be calculated on a Turing machine" (Pagels, 1988, p. 294).

Koch provided a more specific, but still generic, definition: a computing system maps a physical system (such as beads on an abacus) to a more abstract system (such as the natural numbers); then—when presented with data—they respond on the basis of the representations according to some algorithm, usually transforming the original representations. By this definition the brain, indubitably, is a computer. The world is represented to us through the medium of our senses, and, through a series of physical events in the nervous system, brain states emerge that we take to be a representation of the world. That these states are not identical with the world is manifest, the further up the ladder of higher thought we climb. Thus there is clearly a transformation. Whether the algorithm that describes how it comes about will ever be fully specified remains for the future of neuroscience. That the brain is performing what Koch calls "the biophysics of computation," however, seems indisputable.

Computational neuroscience has developed such a momentum of intriguing and impressive insights and models that the formerly lively debate (Churchland and Churchland, 1990; Searle, 1990) over whether or not the brain is a computer is beginning to seem somewhat academic and sterile. Metaphorically and literally, computers represent, process, and store information. The brain does all of these things, and with a skill and overall speed that almost always surpasses even the most powerful computers produced thus far.

Koch has phrased it more dramatically: "Over the past 600 million years, biology has solved the problem of processing massive amounts of noisy and highly redundant information in a constantly changing environment by evolving networks of billions of highly interconnected nerve cells." Because these information processing systems have evolved, Koch challenged, "it is the task of scientists to understand the principles underlying information processing in these complex structures." The pursuit of this challenge, as he suggested, necessarily involves multidisciplinary perspectives not hitherto embraced cleanly under the field of neuroscience. "A new field is emerging," he continued: "the study of how computations can be carried out in extensive networks of heavily interconnected processing elements, whether these networks are carbon- or silicon-based."

NEURAL NETWORKS OF THE BRAIN

In the 19th century, anatomists looking at the human brain were struck with its complexity (the common 20th-century metaphor is its wiring). Some of the most exquisite anatomical drawings were those by Spaniard anatomist Ramón y Cajal, whose use of Golgi staining revealed under the microscope the intricate branching and myriad connections of the brain's constituent parts. These elongated cells have been called neurons from the time of Aristotle and Galen. Throughout the history of neuroscience, the spotlight has swung back and forth, onto and away from these connections and the network they constitute. The question is how instrumental the wiring is to the brain states that develop. On one level, for decades, neuroscientists have studied the neuronal networks of sea slugs and worms up through cats and monkeys. This sort of analysis often leads to traditional, biological hypothesizing about the exigencies of adaptation, survival, and evolutionary success. But in humans, where neuronal networks are more difficult to isolate and study, another level of question must be addressed. Could Mozart's music, Galileo's insight, and Einstein's genius, to mention but a few examples—be explained by brains "wired together" in ways subtly but importantly different than in other humans?

The newest emphasis on this stupendously elaborate tapestry of nerve junctures is often referred to as connectionism, distinguished by its credo that deciphering the wiring diagram of the brain, and understanding what biophysical computations are accomplished at the junctions, should lead to dramatic insights into how the overall

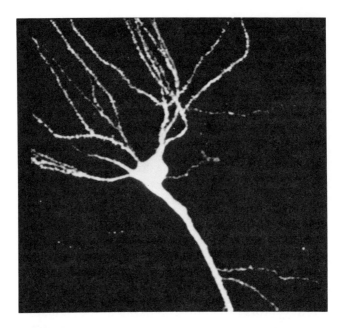

FIGURE 9.1 Nerve cell from the hippocampus, reconstituted by confocal microscopy. (Courtesy of P. Adams.)

system works (Hinton and Anderson, 1981). How elaborate is it? Estimates place the number of neurons in the central nervous system at between 10^{10} and 10^{11}; on average, said Koch, "each cell in the brain is connected with between 1,000 and 10,000 others." A good working estimate of the number of these connections, called synapses, is a hundred million million, or 10^{14}.

Adams, a biophysicist, designs models and conducts experiments to explore the details of how the basic electrical currency of the brain is minted in each individual neuron (Figure 9.1). He has no qualms referring to it as a bottom-up approach, since it has become highly relevant to computational neuroscience ever since it became appreciated "that neurons do not function merely as simple logical units" or on–off switches as in a digital computer. Rather, he explained, a large number of biophysical mechanisms interact in a highly complex manner that can be thought of as computing a nonlinear function of a neuron's inputs. Furthermore, depending on the state of the cell, the same neuron can compute several quite different functions of its inputs. The "state" Adams referred to can be regulated both by chemical modulators impinging on the cell as well as by the cell's past history.

To demonstrate another catalog of reasons why the brain-as-digital-computer metaphor breaks down, Adams described three types of electrical activity that distinguish nerve cells and their connections from passive wires. One type of activity, synaptic activity, occurs at the connections between neurons. A second type, called the action potential, represents a discrete logical pulse that travels in a kind of chain reaction along a linear path down the cell's axon (see Box 9.1). The third type of activity, subthreshold activity, provides the links between the other two.

Each type of electrical activity is caused by "special protein molecules, called ion channels, scattered throughout the membrane of the nerve cells," said Adams. These molecules act as a sort of tiny molecular faucet, he explained, which, when turned on, "allows a stream of ions to enter or leave the cell (Figure 9.2). The molecular faucet causing the action potential is a sodium channel. When this channel is open, sodium streams into the cell," making its voltage positive. This voltage change causes sodium faucets further down the line to open in turn, leading to a positive voltage pulse that travels along the neuron's axon to the synapses that represent that cell's connections to other neurons in the brain. This view of "the action potential," he said, "was established 30 years ago and is the best-founded theory in neurobiology."

The other types of activity, synaptic and subthreshold, "are more complex and less well understood," he said, but it is clear that they, too, involve ion channels. These ion channels are of two types, which either allow calcium to stream into the cell or allow potassium to leave it. Adams has explored the mechanisms in both types of channel. Calcium entering a neuron can trigger a variety of important effects, causing chemical transmitters to be released to the synapses, or triggering potassium channels to open that initiate other subthreshold activity, or producing the long-term changes at synapses that underlie memory and learning.

The subcellular processes that govern whether—and how rapidly—a neuron will fire occur on a time scale of milliseconds but can be visualized using a scanning laser microscope. Activated by additional channels that allow double-charged calcium ions (Ca^{++}) to rapidly enter the cell are two types of potassium channels. One of these channels quickly terminates the sodium inrush that is fueling the action potential. The other one responds much more slowly, making it more difficult for the cell to fire spikes in quick succession. Yet other types of potassium channels are triggered to open, not by calcium, but rather in response to subthreshold voltage changes. Called "M" and "D," these channels can either delay or temporarily prevent

BOX 9.1   HOW THE BRAIN TRANSMITS ITS SIGNALS

The Source and Circuitry of the Brain's Electricity

Where does the brain's network of neurons come from? Is it specified by DNA, or does it grow into a unique albeit human pattern that depends on each person's early experiences? These alternative ways of framing the question echo a debate that has raged throughout the philosophical history of science since early in the 19th century. Often labelled nature vs. nurture, the controversy finds perhaps its most vibrant platform in considering the human brain. How much the DNA can be responsible for the brain's wiring has an inherent limit. Since there are many more synaptic connections—"choice points"—in a human brain network during its early development than there are genes on human chromosomes, at the very least genes for wiring must specify not synapse-by-synapse connections, but larger patterns. The evolution of the nervous system over millions of years began with but a single cell sensing phenomena near its edge, and progressed to the major step that proved to be the pivotal event leading to the modern human brain—the development of the cerebral cortex.

Koch described the cortex as a "highly convoluted, 2-millimeter-thick sheet of neurons. If unfolded, it would be roughly the size of a medium pizza, about 1500 square centimeters." It is the cerebral cortex that separates the higher primates from the rest of living creatures, because most thinking involves the associative areas located in its major lobes. A cross section of this sheet reveals up to six layers, in each of which certain types of neurons predominate. The density of neurons per square millimeter of cortical surface Koch put at about 100,000. "These are the individual computational elements, fundamentally the stuff we think with," he explained.

Although there are many different types of neurons, they all function similarly. And while certain areas of the brain and certain networks of neurons have been identified with particular functions, it is the basic activity of each of the brain's 10 billion neurons that constitutes the working brain. Neurons either fire or they do not, and an electrical circuit is thereby completed or interrupted. Thus, how a single neuron works constitutes the basic foundation of neuroscience.

The cell body (or soma) seems to be the ultimate destination of the brain's electrical signals, although the signals first arrive at the thousands of dendrites that branch off from each cell's soma. The nerve cell sends its signals through a different branch, called an axon. Only one main axon comes from each cell body, although it may branch many times into collaterals in order to reach all of the cells to which it is connected for the purpose of sending its electrical signal.

BOX 9.1—*continued*

Electricity throughout the nervous system flows in only one direction: when a cell fires its "spike" of electricity, it is said to emit an action potential that travels at speeds up to about 100 miles per hour. A small electric current begins at the soma and travels down the axon, branching without any loss of signal as the axon branches. The axon that is carrying the electric charge can run directly to a target cell's soma, but most often it is the dendrites branching from the target cell that meet the transmitting axon and pick up its signal. These dendrites then communicate the signal down to their own soma, and a computation inside the target cell takes place. Only when the sum of the inputs from the hundreds or thousands of the target cell's dendrites achieves a threshold inherent to that particular cell will the cell fire. These inputs, or spikes, are bursts of current that vary only in their pulse frequency, not their amplitude.

Each time a given nerve cell fires, its signal is transmitted down its axon to the same array of cells. Every one of the 10^{15} connections in the nervous system is known as a synapse, whether joining axon to dendrite or axon to cell body. One final generalization about the brain's electricity: when a cell fires and transmits its characteristic (amplitude) electrical message, it may vary in two ways. First, it can emit a more frequent pulse. Second, when the action potential reaches the synaptic connections that the axon makes with other cells, it can either excite the target cell's electrical activity or inhibit it. Every cell has its own inherent threshold, which must be achieved in order for the cell to spike and send its message to all of the other cells to which it is connected. Whether that threshold is reached depends on the cumulative summation of all of the electrical signals—both excitatory and inhibitory—that a cell receives in a finite amount of time.

Simple Chemistry Provides the Brain's Electricity

The fluid surrounding a neuron resembles dilute seawater, with an abundance of sodium and chloride ions, and dashes of calcium and magnesium. The cytoplasm inside the cell is rich in potassium ions, and many charged organic molecules. The neuron membrane is a selective barrier separating these two milieu. In the resting state only potassium can trickle through, which as it escapes creates a negative charge on the cytoplasm. The potential energy represented by this difference in polarity for each cell is referred to as its resting potential and can be measured at about a tenth of a volt. Twenty such cells match the electrical energy in a size-D flashlight battery.

On closer inspection, the cell membrane actually consists of myriad channels, each of which has a molecular structure configured to

continued on next page

BOX 9.1—*continued*

let only one particular ion substance pass through its gate. When the nerve is in its resting state, most of these gates are closed. The nerve is triggered to fire when a critical number of its excitatory synapses receive neurotransmitter signals that tip the balance of the cell's interior charge, causing the gates on the sodium channels to open. Sodium ions near these gates rush in for simple chemical reasons, as their single positive charge left by the missing outer-shell electron is attracted to the negative ions inside the cell. This event always begins near the cell body. However, as soon as the inrushing sodium ions make the polarity more positive in that local region, nearby sodium gates further down the axon are likewise triggered to open.

This process of depolarization, which constitutes the electrical action potential, is self-perpetuating. No matter how long the journey down the axon—and in some nerve cells it can be more than a meter—the signal strength is maintained. Even when a damaged region of the cell interrupts the chemical chain reaction, the signal resumes its full power as it continues down the axon beyond the compromised region. The duration of the event is only about 0.001 second, as the system quickly returns to equilibrium.

The foregoing sketch of the chemistry explains the excitatory phase and why cells do fire. Conversely, many synapses are inhibitory and instead of decreasing the cell's negative polarity actually increase it, in part by opening chloride gates. This explains, in part, why cells do not fire.

This picture of how neurons fire leads to the explanation of why a given cell will begin the action-potential chain reaction. Only when its polarity changes sufficiently to open enough of the right kind of gates to depolarize it further will the process continue. The necessary threshold is achieved by a summation of impulses, which occur for either or both of two reasons: either the sending neuron fires rapidly in succession or a sufficiently large number of excitatory neurons are firing. The result is the same in either case: when the charge inside the cell decreases to a certain characteristic value, the sodium gates start to open and the neuron depolarizes further. Once begun, this sequence continues throughout the axon and all of its branches, and thereby transforms the receiving nerve cell into a transmitting one, whose inherent electrical signal continues to all of the other cells connected to it. Thus, from one neuron to another, a signal is transmitted and the path of the electricity defines a circuit.

SOURCES: Changeux, 1986, pp. 80ff.; Bloom and Lazerson, 1988, pp. 36–40.

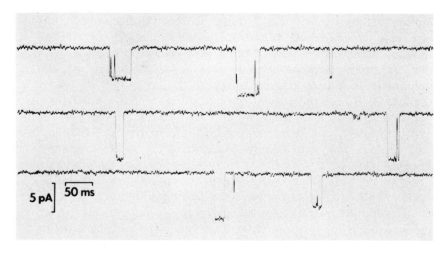

FIGURE 9.2 Single ion channels in a membrane opening and closing. (Courtesy of P. Adams.)

the cell's firing that active synapses would otherwise cause. Adams and his colleagues have shown that many of these calcium and potassium channels are regulated by "neuromodulators," chemicals released by active synapses.

Working energetically from the bottom up, "the cellular or molecular neurobiologist may appear to be preoccupied by irrelevant details" that shed no light on high-level functions, Adams conceded. "Knowing the composition of the varnish on a Stradivarius is irrelevant to the beauty of a chaconne that is played with it. Nevertheless," he asserted, "it is clear that unless such details are just right, disaster will ensue." Any more global theory of the brain or the higher or emergent functions it expresses must encompass theories that capture these lower-level processes. Adams concluded by giving "our current view of the brain" as seen from the perspective of an experimental biophysicist: "a massively parallel analog electrochemical computer, implementing algorithms through biophysical processes at the ion-channel level."

FROM COMPUTERS TO ARTIFICIAL INTELLIGENCE TO CONNECTIONISM

Models—Attempts to Enhance Understanding

The complexity of the brain's structure and its neurochemical firing could well stop a neural net modeler in his or her tracks. If the factors influencing whether just one neuron will fire are so complex, how can a model remain "faithful to the biological constraints," as many neuroscientists insist it must, and yet still prove useful? This very pragmatic question actually provides a lens to examine the development of neuroscience throughout the computational era since the late 1940s, for the advent of computers ignited a heated controversy fueled by the similarities—and the differences—between brains and computers, between thinking and information processing. Adams believes that neuroscience is destined for such skirmishes and growing pains because—relative to the other sciences—it is very young. "Really it is only since World War II that people have had any real notion about what goes on there at all," he pointed out.

Throughout this history, an underlying issue persists: what is a model, and what purpose does it serve? Much of the criticism of computer models throughout the computational era has included the general complaint that models do not reflect the brain's complexity fairly, and current neural networkers are sensitive to this issue. "It remains an interesting question," said Ullman, "whether your model represents a realistic simplification or whether you might really be throwing away the essential features for the sake of a simplified model." Koch and his colleague Idan Segev, however, preferred to reframe the question, addressing not whether it is desirable to simplify, but whether it is possible to avoid doing so (they say it is not) and—even if it were—to ask what useful purpose it would serve. Their conclusion was that "even with the remarkable increase in computer power during the last decade, modeling networks of neurons that incorporate all the known biophysical details is out of the question. Simplifying assumptions are unavoidable" (Koch and Segev, 1989, p. 2).

They believe that simplifications are the essence of all scientific models. "If modeling is a strategy developed precisely because the brain's complexity is an obstacle to understanding the principles governing brain function, then an argument insisting on complete biological fidelity is self-defeating," they maintained. In particular, the brain's complexity beggars duplication in a model, even while it begs for a metaphorical, analogical treatment that might yield further scientific insight. (Metaphors have laced the history of neuroscience, from Sherrington's enchanted loom to British philosopher Gilbert Ryle's ghost in the machine, a construct often employed in the effort to embody emergent properties of mind by those little concerned with the corporeal nature of the ghost that produced them.) But Koch and his colleagues "hope that realistic models of the cerebral cortex do not need to incorporate explicit information about such minutiae as the various ionic channels [that Adams is studying]. The style of simplifying assumptions they suggest could be useful "[might be analogous to] . . . Boyle's gas law, where no mention is made of the 10^{23} or so molecules making up the gas." Physics, chemistry, and other sciences concerned with minute levels of quantum detail, or with the location of planets and stars in the universe, have experienced the benefit of simplified, yet still accurate, theories.

Writing with Sejnowski and the Canadian neurophilosopher Patricia Churchland, Koch has made the point that scientists have already imposed a certain intellectual structure onto the study of the brain, one common in scientific inquiry: analysis at different levels (Sejnowski et al., 1988). In neuroscience, this has meant the study of systems as small as the ionic channels (on the molecular scale, measured in nanometers) or as large as working systems made up of networks and the maps they contain (on the systems level, which may extend over 10 centimeters). Spanning this many orders of magnitude requires that the modeler adapt his or her viewpoint to the scale in question. Evidence suggests that the algorithms that are uncovered are likely to bridge adjacent levels. Further, an algorithm may be driven by the modeler's approach. Implementing a specific computational task at a particular level is not necessarily the same as asking what function is performed at that level. The former focuses on the physical substrate that is accomplishing the computation, the latter on what functional role the system is accomplishing at that level. In either case, one can look "up" toward emergent brain states, or "down" toward the biophysics of the working brain. Regardless of level or viewpoint, it is the computer—both as a tool and as a concept—that often enables the modeler to proceed.

Computation and Computers

Shortly after the early computers began to crunch their arrays of numbers and Turing's insights began to provoke theorists, Norbert Wiener's book *Cybernetics* was published, and the debate was joined over minds vs. machines, and the respective province of each. Most of the current neural net modelers developed their outlooks under the paradigm suggested above; to wit, the brain's complexity is best approached not by searching for some abstract unifying algorithm that will provide a comprehensive theory, but rather by devising models inspired by how the brain works—in particular, how it is wired together, and what happens at the synapses. Pagels included this movement, called connectionism, among his new "sciences of complexity": scientific fields that he believed use the computer to view and model the world in revolutionary and essential ways (Pagels,

1988). He dated the schism between what he called computational-ists and connectionists back to the advent of the computer (though these terms have only assumed their current meaning within the last decade).

At the bottom of any given revival of this debate was often a core belief as to whether the brain's structure was essential to thought—whether, as the connectionists believed, intelligence is a property of the design of a network. Computationalists wanted to believe that their increasingly powerful serial computers could manipulate symbols with a dexterity and subtlety that would be indistinguishable from those associated with humans, hence the appeal of the Turing test for intelligence and the birth of the term (and the field) *artificial intelligence*. A seminal paper by Warren McCulloch and Walter H. Pitts in 1943 demonstrated that networks of simple McCullough and Pitts neurons are Turing-universal: anything that can be computed, can be computed with such networks (Allman, 1989). Canadian neuroscientist Donald Hebb in 1949 produced a major study on learning and memory that suggested neurons in the brain actually change—strengthening through repeated use—and therefore a network configuration could "learn," that is, be enhanced for future use.

Throughout the 1950s the competition for funding and converts continued between those who thought fidelity to the brain's architecture was essential for successful neural net models and those who believed artificial intelligence need not be so shackled. In 1962, in Frank Rosenblatt's *Principles of Neurodynamics*, a neural net model conceptualizing something called perceptrons curried much favor and attention (Allman, 1989). When Marvin Minsky and Seymour Papert published a repudiation of perceptrons in 1969, the artificial intelligence community forged ahead of the neural net modelers. The Turing machine idea empowered the development of increasingly abstract models of thought, where the black box of the mind was considered more a badge of honor than a concession to ignorance of the functioning nervous system. It was nearly a decade later before it was widely recognized that the valid criticism of perceptrons did not generalize to more sophisticated neural net models. Meanwhile one of Koch's colleagues, computer scientist Tomaso Poggio, at the Massachusetts Institute of Technology, was collaborating with David Marr to develop a very compelling model of stereoscopic depth perception that was clearly connectionist in spirit. "Marr died about 10 years ago," said Koch, but left a strong legacy for the field. Although he was advocating the top-down approach, his procedures were "so tied to biology" that he helped to establish what was to become a more unified approach.

Turning Top-down and Bottom-up Inside Out

Although always concerned about the details of neurobiology, Marr was strongly influenced by the theory of computation as it developed during the heyday of artificial intelligence. According to Churchland et al. (1990), Marr characterized three levels: (1) the computational level of abstract problem analysis . . . ; (2) the level of the algorithm . . . ; and (3) the level of physical implementation of the computation" (pp. 50–51). Marr's approach represented an improvement on a pure artifical intelligence approach, but his theory has not proven durable as a way of analyzing how the brain functions. As Sejnowski and Churchland have pointed out, "'When we measure Marr's three *levels of analysis* against *levels of organization* in the nervous system, the fit is poor and confusing'" (quoted in Churchland et al., 1990, p. 52). His scheme can be seen as another in a long line of conceptualizations too literally influenced by a too limited view of the computer. As a metaphor, it provides a convenient means of classification. As a way to understand the brain, it fails.

Inadequacy of the Computer as a Blueprint for How the Brain Works

Nobelist and pioneering molecular biologist Francis Crick, together with Koch, has written about how this limited view has constrained the development of more fruitful models (Crick and Koch, 1990). "A major handicap is the pernicious influence of the paradigm of the von Neumann digital computer. It is a painful business to try to translate the various boxes and labels, 'files,' 'CPU,' 'character buffer,' and so on occurring in psychological models—each with its special methods of processing—into the language of neuronal activity and interaction," they pointed out. In the history of thinking about the mind, metaphors have often been useful, but at times misleadingly. "The analogy between the brain and a serial digital computer is an exceedingly poor one in most salient respects, and the failures of similarity between digital computers and nervous systems are striking," wrote Churchland et al. (1990, p. 47).

Some of the distinctions between brains and computers: the von Neumann computer relies on an enormous landscape of nonetheless very specific places, called addresses, where data (memory) is stored. The computer has its own version of a "brain," a central processing unit (CPU) where the programmer's instructions are codified and carried out. Since the entire process of programming relies on logic, as does the architecture of the circuitry where the addresses can be found, the basic process generic to most digital computers can be described as a series of exactly specified steps. In fact this is the definition of a computer algorithm, according to Churchland. "You take this and do that. Then take that and do this, and so on. This is exactly what computers happen to be great at," said Koch. The computer's CPU uses a set of logical instructions to reply to any problem it was programmed to anticipate, essentially mapping out a step-by-step path through the myriad addresses to where the desired answer ultimately resides. The address for a particular answer usually is created by a series of operations or steps that the algorithm specifies, and some part of the algorithm further provides instructions to retrieve it for the microprocessor to work with. A von Neumann computer may itself be specified by an algorithm—as Turing proved—but its great use as a scientific tool resides in its powerful hardware.

Scientists trying to model perception eschew the serial computer as a blueprint, favoring instead network designs more like that of the brain itself. Perhaps the most significant difference between the two is how literal and precise the serial computer is. By its very definition, that literalness forbids the solution of problems where no precise answer exists. Sejnowski has collaborated on ideas and model networks with many of the leaders in the new movement, among them Churchland. A practicing academician and philosopher, Churchland decided to attend medical school to better appreciate the misgivings working neuroscientists have developed about theories that seem to leave the functioning anatomy behind. In her book *Neurophilosophy: Toward a Unified Science of the Mind/Brain*, she recognized a connection between neural networks and the mind: "'Our decision making is much more complicated and messy and sophisticated—and powerful—than logic . . . [and] may turn out to be much more like the way neural networks function: The neurons in the network interact with each other, and the system as a whole evolves to an answer. Then, introspectively, we say to ourselves: I've decided'" (Allman, 1989, p. 55).

The Computer as a Tool for Modeling Neural Systems

Critiques like Crick's of the misleading influence the serial computer has had on thinking about the brain do not generalize about other types of computers, nor about the usefulness of the von Neumann machines as a tool for neural networkers. Koch and Segev wrote that in fact "computers are the *conditio sine qua non* for studying the behavior of the model[s being developed] for all but the most trivial cases" (Koch and Segev, 1989, p. 3). They elaborated on the theme raised by Pagels, about the changes computation has wrought in the very fabric of modern science: "Computational neuroscience shares this total dependence on computers with a number of other fields, such as hydrodynamics, high-energy physics, and meteorology, to name but a few. In fact, over the last decades we have witnessed a profound change in the nature of the scientific enterprise" (p. 3). The tradition in Western science, they explained, has been a cycle that runs from hypothesis to prediction, to experimental test and analysis, a cycle that is repeated over and again and that has led, for example, "to the spectacular successes of physics and astronomy. The best theories, for instance Maxwell's equations in electrodynamics, have been founded on simple principles that can be relatively simply expressed and solved" (p. 3).

"However, the major scientific problems facing us today probably do not have theories expressible in easy-to-solve equations," they asserted (p. 3). "Deriving the properties of the proton from quantum chromodynamics, solving Schroedinger's equation for anything other than the hydrogen molecule, predicting the three-dimensional shape of proteins from their amino-acid sequence or the temperature of earth's atmosphere in the face of increasing carbon dioxide levels— or understanding the brain—are all instances of theories requiring massive computer power" (p. 3). As firmly stated by Pagels in his treatise, and as echoed throughout most of the sessions at the Frontiers symposium, "the traditional Baconian dyad of theory and experiment must be modified to include *computation*," said Koch and Segev. "Computations are required for making predictions as well as for evaluating the data. This new triad of theory, computation and experiment leads in turn to a new [methodological] cycle: mathematical theory, simulation of theory and prediction, experimental test, and analysis" (p. 3).

Accounting for Complexity in Constructing Successful Models

Another of the traditional ways of thinking about brain science has been influenced by the conceptual advances now embodied in computational neuroscience. Earlier the brain–mind debate was framed in terms of the sorts of evidence each side preferred to use. The psychologists trying to stake out the terrain of "the mind" took an approach that was necessarily top-down, dominated by a view known as general problem solving (GPS), by which they meant that, as in a Turing machine, the software possessed the answers, and what they sometimes referred to as the "wetware" was largely irrelevant, if not immaterial. This did not prevent their constructing models, however, said Koch. "If you open a book on perception, you see things like 'input buffer,' 'CPU,' and the like, and they use language that suggests you see an object in the real world and open a computer file in the brain to identify it."

Neurophysiologists in particular have preferred to attend to the details and complexities rather than grasping at theories and patterns that sometimes seem to sail off into wonderland. "As a neuroscientist I have to ask," said Koch, "Where is this input buffer? What does it mean to 'open a file in the brain?'"

Biological systems are quintessentially dynamic, under the paradigm of neo-Darwinian evolution. Evolution designed the brain one step at a time, as it were. At each random, morphological change, the test of survival was met not by an overall plan submitted to conscious introspection, but rather by an organism that had evolved a specific set of traits. "Evolution cannot start from scratch," wrote Churchland et al. (1990), "even when the optimal design would require that course. As Jacob [1982] has remarked, evolution is a tinkerer, and it fashions its modification out of available materials, limited by earlier decisions. Moreover, any given capacity . . . may look like a wonderfully smart design, but in fact it may not integrate at all well with the wider system and may be incompatible with the general design of the nervous system" (p. 47). Thus, they conclude, in agreement with Adams, "neurobiological constraints have come to be recognized as essential to the project" (p. 47).

If a brain does possess something analogous to computer programs, they are clearly embedded in the scheme of connections and within the billions of cells of which that brain is composed, the same place where its memory must be stored. That place is a terrain that Adams has explored in molecular detail. "I really don't believe that we're dealing with a software–hardware relationship, where the brain possesses a transparent program that will work equally well on any type of computer. There's such an intimate connection," he stressed, "between *how* the brain does the computation and the computations that it does, that you can't understand one without the other." Almost all modern neuroscientists concur.

The Parallel, Distributed Route to New Insights

By the time Marr died in 1981, artifical intelligence had failed to deliver very compelling models about global brain function (in general, and memory in particular), and two essential conclusions were becoming apparent. First, the brain is nothing like a serial computer, in structure or function. Second, since memory could not be assigned a specific location, it must be distributed throughout the brain and must bear some important relationship to the networks through which it was apparently laid down and retrieved.

It was these and other distinctions between brains and serial computers that led physicist John Hopfield to publish what was to prove a seminal paper (Hopfield, 1982). The Hopfield network is, in a sense, the prototypical neural net because of its mathematical purity and simplicity, but still reflects Hopfield's long years of studying the brain's real networks. "I was trying to create something that had the essence of neurobiology. My network is rich enough to be interesting, yet simple enough so that there are some mathematical handles on it," he wrote (p. 81). Koch agreed: "There were other neural networks around similar to Hopfield's, but his perspective as a physicist" turned out to be pivotal. Hopfield described his network, said Koch, "in a very elegant manner, in the language of physics. He used clear and meaningful terms, like 'state space,' and 'variational minimum,' and 'Lyapunov function.' Hopfield's network was very elegant and simple, with the irrelevancies blown away." Its effect was galvanic, and suddenly neural networks began to experience a new respectability.

David Rumelhart of Stanford University, according to Koch, "was the guiding influence" behind what was to prove a seminal collection of insights that appeared in 1986. *Parallel Distributed Processing: Explorations in the Microstructure of Cognition* (Rumelhart and McClelland, 1986) included the work of a score of contributing authors styled as the PDP research group, among whose number were Crick and Sejnowski. As the editors said in their introduction, "One of the great joys of science lies in the moment of shared discovery. One person's half-baked suggestion resonates in the mind of another and suddenly takes on a definite shape. An insightful critique of one way of thinking about a problem leads to another, better understanding. An incomprehensible simulation result suddenly makes sense as two people try to understand it together. This book grew out of many such moments" (p. ix). For years, various combinations of these people, later augmented by Churchland and others, have met regularly at the University of California, San Diego, where Sejnowski now works.

This two-volume collection contains virtually all of the major ideas then current about what neural networks can and might do. The field is progressing rapidly, and new ideas and models continue to appear in journals such as *Network* and *Neural Computation*. As a generalization, neural networks are designed to try to capture some of the abilities brains demonstrate when they think, rather than the logic-based rules and deductive reasoning computers embody when they process information. The goal of much work in neural networks is to develop one or more of the abilities usually believed to require human judgment and insight, for example, the abilities to recognize patterns, make inferences, form categories, arrive at a consensus "best guess" among numerous choices and competing influences, forget when necessary and appropriate, learn from mistakes, learn by training, modify decisions and derive meaning according to context, match patterns, and make connections from imperfect data.

VISION MODELS OPEN A WINDOW ON THE BRAIN

Koch's presentation on unravelling the complexities of vision suggested the basic experimental approach in computational neuroscience: Questions are posed about brain states and functions, and the answers are sought by pursuing a continually criss-crossing path from brain anatomy to network behavior to mathematical algorithm to computer model and back again.

One begins to narrow to a target by developing an algorithm and a physical model to implement it, and by testing its results against the brain's actual performance. The model may then be refined until it appears able to match the real brains that are put to the same experimental tests. If a model is fully specified by one's network, the model's structure then assumes greater significance as a fortified guess about how the brain works. The real promise of neural networks is manifest when a model is established to adjust to and learn from the experimental inputs and the environment; one then examines how the network physically modified itself and "learned" to compute competitively with the benchmark brain.

Motion Detectors Deep in the Brain

As Koch explained to the symposium scientists: "All animals with eyes can detect at least some rudimentary aspects of visual motion. . . . Although this seems a simple enough operation to understand, it turns out to be an enormously difficult problem: What are the computations underlying this operation, and how are they implemented in the brain?" Specifically, in the case of motion, the aim is to compute the optical flow field, that is, to "attach" to every moving location in the image a vector whose length is proportional to the speed of the point and whose direction corresponds to the direction of the moving object (Figure 9.3). As used here, the term *computation* carries two meanings. First, the network of neuronal circuitry projecting from the eye back into the brain can be viewed as having an input in the form of light rays projecting an image on the retina, and an output in terms of perception—what the brain makes of the input. Human subjects can report on many of their brain states, but experiments with other species raise the problem of the scientist having to interpret, rather than merely record, the data. Koch models the behavior of a species of monkey, *Macaca mulatta*, which has a visual system whose structure is very close to that found in humans and which is very susceptible to training. Although it is not possible to inquire directly of the animal what it sees, some very strong inferences can be made from its behavior. The second sense of computation involves the extent to which the input–output circuit Koch described can be thought of as instantiating a computation, i.e., an information-processing task.

Before a simulation progresses to the stage that an analog network prototype is physically constructed and experiments are run to see how it will perform, the scientist searches for an algorithm that seems to capture the observable data. "Computational theory provides two broad classes of algorithms computing the optical flow associated with changing image brightness," said Koch, who has chosen and refined the one he believes best explains the experimental data. He established some of the basic neuroanatomy for the symposium's scientists to provide a physical context in which to explain his model.

In the monkey's brain nearly 50 percent of the cortex (more than in the human brain) is related to vision, and about 15 percent of that constitutes the primary visual area (V1) at the back of the head, the first place, in moving through the monkey brain's visual circuitry, where cells are encountered that seem inherently sensitive to the direction of a perceived motion in the environment. "In connectionists' parlance, they are called *place* or *unit* coded, because they respond to motion only in a particular—their preferred—direction," Koch explained. At the next step in the circuit, the middle temporal area (MT), visual information clearly undergoes analysis for motion cues, since, he said, "80 percent of all cells in the MT are directionally selective and tuned for the speed of the stimulus."

One hurdle to overcome for effectively computing optical flow is the so-called aperture problem, which arises because each cortical cell only "sees" a very limited part of the visual field, and thus "only the component of velocity normal to the contour can be measured, while the tangential component is invisible," Koch said. This problem is inherent to any computation of optical flow and essentially presents the inquirer with infinitely many solutions. Research into

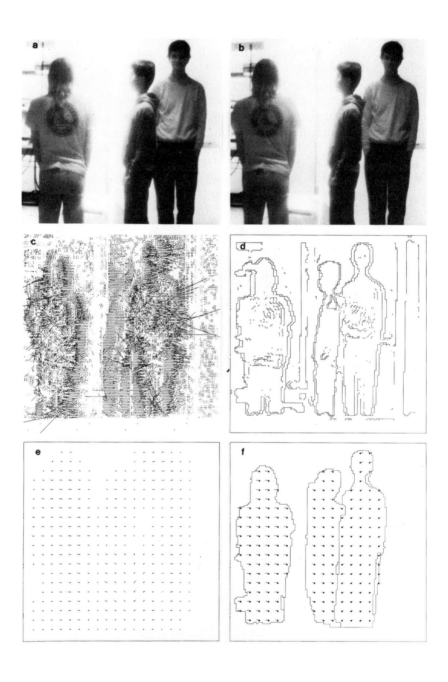

vision has produced a solution in the form of the smoothness constraint, "which follows not from a mathematical analysis of the problem, but derives rather from the physics of the real world," Koch said, since most objects in motion move smoothly along their smooth trajectory (i.e., they do not generally stop, speed up, or slow down) and nearby points on the same object move roughly at the same velocity. "In fact, the smoothness constraint is something that can be learned from the environment," according to Koch, and therefore might well be represented in the visual processing system of the brain, when it is considered as a product of evolution. Ullman clarified that the smoothness assumption "concerns smoothness in space, not in time, that nearby points on the same object have similar velocities."

Equipped with these assumptions and modeling tools—the aperture problem, the smoothness constraint, and a generic family of motion algorithms—Koch and his colleagues developed what he described as a "two-stage recipe for computing optical flow, involving a local motion registration stage, followed by a more global integration stage." His first major task was to duplicate "an elegant psychophysical and electrophysiological experiment" that supported the breakdown of visual processing for motion into these two stages and established a baseline against which he could measure the value of his own model. To summarize that experiment (Adelson and Movshon, 1982): When two square gratings were projected in motion onto a screen under

---

FIGURE 9.3 Optical flow associated with several moving people. (a) and (b) Two 128- by 128-pixel images captured by a video camera. The two people on the left move toward the left, while the rightward person moves toward the right. (c) Initial velocity data. Since each "neuron" sees only a limited part of the visual scene, the initial output of these local motion sensors is highly variable from one location to the next. (d) Threshold edges (zero-crossings of the Laplacian of a Gaussian) associated with one of the images. (e) The smooth optical flow after the initial velocity measurements (in c) have been integrated and averaged (smoothed) many times. Notice that the flow field is zero between the two rightmost people, since the use of the smoothness constraint acts similarly to a spatial average and the two opposing local motion components average out. The velocity field is indicated by small needles at each point, whose amplitude indicates the speed and whose direction indicates the direction of motion at that point. (f) The final piecewise smooth optical flow after 13 complete cycles shows a dramatic improvement in the optical flow. Motion discontinuities are constrained to be co-localized with intensity edges. The velocity field is subsampled to improve visibility. (Reprinted with modifications from Hutchinson et al., 1988.)

certain provisos, human test subjects (as well as monkeys trained to indicate what they saw) failed to see them independently, but rather integrated them in a way that suggested the brain was computing normal components of motion into a coherent, unified pattern. "Under certain conditions you see these two gratings penetrate past each other. But under the right conditions," said Koch, "you don't see them moving independently. Rather you see one moving plaid," suggesting that the brain was computing normal components of motion received by the retina into a coherent unified pattern. While this perception may be of use from evolution's point of view, it does not reflect reality accurately at what might be called, from a computational perspective, the level of raw data. The monkeys performed while electrodes in their brains confirmed what was strongly suspected: the presence of motion-detecting neurons in the V1 and MT cortex regions.

It was in this context that Koch and his colleagues applied their own neural network algorithm for computing motion. Specifically, they tackled the question of why "any one particular cell acquires direction selectivity, such that it only fires for motion in its preferred direction, not responding to motion in most other directions," Koch explained. Among other features, their model employed population coding, which essentially summates all of the neurons under examination. Koch, often influenced by his respect for evolution's brain, believes this is important because while "such a representation is expensive in terms of cells, it is very robust to noise or to the deletion of individual cells, an important concern in the brain," which loses thousands of neurons a day to attrition. The network, when presented with the same essential input data as were the test subjects, mimicked the behavior of the real brains.

To further test the correspondence of their network model to how real brains detect motion, Koch and his colleagues decided to exploit the class of perceptual phenomena called illusions (defined by Koch as those circumstances when "our perception does not correspond to what is really out there"). Motion capture is one such illusion that probably involves visual phenomena related to those they were interested in. In an experiment generally interpreted as demonstrating that the smoothness constraint—or something very like it—is operating in the brains of test subjects, Koch's model, with its designed constraint, was "fooled" in the same way as were the test subjects. Even when illusions were tested that appeared not to invoke the smoothness constraint, the model mimicked the biological brain. Yet another experiment involved exposing the subject to a stationary figure, which appeared and disappeared very quickly from the viewing screen. Brains see such a figure appear with the apparent motion of expansion, and disappear with apparent contraction, although "out there" it has remained uniform in size. Koch's network registered the same illusory perception.

Armed with this series of affirmations, Koch asked, "What have we learned from implementing our neuronal network?" Probably, he continued, that the brain does have some sort of procedure for imposing a smoothness constraint, and that it is probably located in the MT area. Koch, cautious about further imposing the mathematical nature of his own algorithm onto a hypothesis about brain function, said, "The exact type of constraint we use is very unlikely" to function exactly like that probably used by the brain. Rather than limiting his inferences, however, this series of experiments reveals "something about the way the brain processes information in general, not only related to motion. In brief," he continued, "I would argue that any reasonable model concerned with biological information processing has to take account of some basic constraints."

"First, there are the anatomical and biophysical limits: one needs to know the structure of V1 and MT, and that action potentials proceed at certain inherent speeds. In order to understand the brain, I must take account of these [physical realities], he continued. But Koch thinks it useful to tease these considerations apart from others that he classifies as "computational constraints. This other consideration says that, in order to do this problem at all, on any machine or nervous system, I need to incorporate something like smoothness." He distinguished between the two by referring to the illusion experiments just described, which—in effect—reveal a world where smoothness does not really apply. However, since our brains are fooled by this anomaly, a neural network that makes the same mistake is probably doing a fairly good job of capturing a smoothness constraint that is essential for computing these sorts of problems. Further, the algorithm has to rapidly converge to the perceived solution, usually in less than a fifth of a second. Finally, another important computational constraint that "brain-like" models need to obey is robustness (to hardware damage, say) and simplicity—the simpler the mathematical series of operations required to find a solution, the more plausible the algorithm, even if the solution is not always correct.

Brain Algorithms Help to Enhance
Vision in a Messy World

For years, he explained, Ullman's quest has been to "understand the visual system," usually taking the form of a question: what computations are performed in solving a particular visual problem? He reinforced Koch's admonition to the symposium's scientists about "how easy it is to underestimate the difficulties and complexities involved in biological computation. This becomes apparent when you try to actually implement some of these computations artificially outside the human brain." It was, he continued, "truly surprising [to find] that even some of the relatively low-level computations . . . like combining the information from the two eyes into binocular vision; color computation; motion computation; edge detection; and so on are all incredibly complicated. We still cannot perform these tasks at a satisfactory level outside the human brain."

Ullman described to the symposium scientists experiments that point to some fairly astounding computational algorithms that the brain, it must be inferred, seems to have mastered over the course of evolution. Once again, the actual speed at which the brain's component parts, the neurons, fire their messages is many orders of magnitude slower than the logic gates in a serial computer. Ullman asked the symposium to think about the following example of the brain overcoming such apparent limitations. Much slower, for example, than the shutter speed of a camera is the 100 milliseconds it takes the brain to integrate a visual image. Ullman elaborated, "Now, think about what happens if you take a photograph of, for example, people moving down the street, with the camera shutter open for 0.1 second. You will get just a complete blur, because even if the image is moving very slowly, during that time it will move 2 degrees and cover at least 25 pixels of the image." Ullman reported that evidence has shown that the image on the surface of the retina is in fact blurred, but the brain at higher levels obviously sees a sharp image.

Ullman cited the implications of another experiment on illusion as illustrating the brain's ingenuity in deriving from its visual input—basically the simple two-dimensional optical flow fields Koch was describing—the physical shape and structure of an object in space. Shading is obviously an important clue to building the necessary perspective for computing a three-dimensional world from the two-dimensional images our retina has to work with. But also crucial is motion. Ullman used the computer to simulate a revealing illusory phenomenon—by painting random dots on two transparent cylinders (same height, different diameter) placed one within the other, and then sending light through the array so it would be projected onto a screen. "The projection of the moving dots was computed and presented on the screen," he continued, and the "sum" of these two random but stationary patterns was itself random, to the human eye. But when Ullman began to rotate the cylinders around their axis, the formerly flat array of dots suddenly acquired depth, and "the full three-dimensional shape of the cylinders became perceivable."

In this experiment, the brain, according to Ullman, is facing the problem of trying to figure out the apparently random motion of the dots. It apparently looks for fixed rigid shapes and, based on general principles, recovers a shape from the changing projection. "The dots are moving in a complex pattern, but can this complex motion be the projection of a simple object, moving rigidly in space? Such an object indeed exists. I now see," said Ullman, speaking for the hypothetical brain, "what its shape and motion are." He emphasized how impressive this feat actually is: "The brain does this on the fly, very rapidly, gives you the results, and that is simply what you see. You don't even realize you have gone through these complex computations."

"'Nature is more ingenious than we are,'" wrote Sejnowski and Churchland recently (Allman, 1989, p. 11). "'The point is, *evolution has already done it*, so why not learn how that stupendous machine, our brain, actually works?'" (p. 11). All of the participants in the session on neural networks conveyed this sense of the brain's majesty and how it far surpasses our ability to fully understand it. Sejnowski reinforced the point as he described his latest collaboration with electrophysiologist Stephen Lisberger from the University of California, San Francisco, in looking at the vestibulo-ocular reflex system: "Steve and many other physiologists and anatomists have worked out most of the wiring diagram, but they still argue about how it works." That is to say, even though the route of synaptic connections from one neuron to another throughout the particular network that contains the reflex is mapped, the organism produces effects that somehow seem more than the sum of its parts. "'We're going to have to broaden our notions of what an explanation is. We will be able to solve these problems, but the solutions won't look like the neat equations we're used to,'" Sejnowski has said (Allman, 1989, p. 89), referring to the emergent properties of the brain that may well be a function of its complexity.

To try to fathom how two of the brain's related but distinct visual subsystems may interact, Sejnowski and Lisberger constructed a neural network to model the smooth tracking and the image stabilization systems (Figure 9.4). "Primates," said Sejnowski, "are particularly good at tracking things with their eyes. But the ability to stabilize images on the retina goes way back to the early vertebrates—if you are a hunter this could be useful information for zeroing in on your prey; if you are the prey, equally useful for noticing there is a predator out there." Like Ullman, he wanted to model the state of an image on the retina and explore how the image is stabilized. The

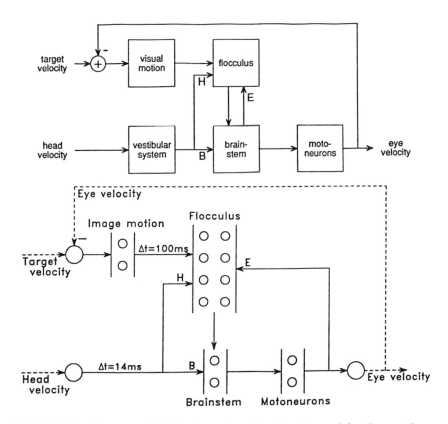

FIGURE 9.4 Diagram of the basic oculomotor circuits used for the vestibu-lo-ocular reflex (VOR) and smooth pursuit (top) and the architecture for a dynamical network model of the system (bottom). The vestibulary system senses head velocity and provides inputs to neurons in the brain stem (B) and the cerebellar flocculus (H). The output neurons of the flocculus make inhibitory synapses on neurons in the brain stem. The output of the brain stem projects directly to motoneurons, which drive the eye muscles, and also project back to the flocculus (E). This positive feedback loop has an "effer-ence copy" of the motor commands. The flocculus also receives visual infor-mation from the retina, delayed for 100 milliseconds by the visual processing in the retina and visual cortex. The image velocity, which is the difference between the target velocity and the eye velocity, is used in a negative feed-back loop to maintain smooth tracking of moving targets. The model (below) incorporates model neurons that have integration time constants and time delays. The arrows indicate connections between the populations of units in each cluster, indicated by small circles. The dotted line between the eye velocity and the summing junction with the target velocity indicates a link-age between the output of the model and the input. (Courtesy of S. Lisberger.)

vestibulo-ocular reflex (VOR) is "incredibly important," according to Sejnowski, "because it stabilizes the object on the retina, and our visual system is very good at analyzing stationary images for meaningful patterns." In the hurly-burly of life, objects do not often remain quietly still at a fixed spot on the retina. An object can move very rapidly or irregularly, and the head in which the visual system is located can also move back and forth, trying to establish a consistent relationship with the object. One of the brain's tricks, said Sejnowski, "is to connect vestibular inputs almost directly to the motor neurons that control where the eye will be directed." Thus, he explained, in only 14 milliseconds those muscles can compensate for a quick head movement with an eye movement in the opposite direction to try to stabilize the image. Somehow the VOR computes this and other reactions to produce a continuous stability of visual input that a camera mounted on our head would never see.

The second brain skill Sejnowski and Lisberger wanted to include in their analysis was the pursuit of visual targets, which involves the interesting feature of feedback information. "In tracking a moving object," Sejnowski explained, "the visual system must provide the oculomotor system with the velocity of the image on the retina." A visual monitoring, feedback, and analysis process can be highly effective but proceeds only at the speed limit with which all of the neurons involved can fire, one after another. The event begins with photons of light hitting the rods and cones in the retina and then being transformed through visual system pathways to cerebral cortex and thence to a part of the cerebellum called the flocculus, and eventually down to the motor neurons, the so-called output whose function is to command the eye's response by moving the necessary muscles to pursue in the right direction. "But there is a problem," Sejnowski said. The "visual system is very slow, and in monkeys takes about 100 milliseconds" to process all the way through, and should therefore produce confusing feedback oscillations during eye tracking. And, he observed, [there is] "a second problem [that] has to do with integrating these two systems, the VOR and the pursuit tracking."

So "our strategy," said Sejnowski, "was to develop a relatively simple network model that could account for all the data, even the controversial data that seem to be contradictory. The point is that even a simple model can help sort out the complexity." Once built, the model could simulate various conditions the brain systems might experience, including having its connection strengths altered to improve performance. Sejnowski then ran it through its paces. He also subjected the model to adaptation experiments that have been performed on humans and monkeys, and unearthed a number of insights about how the brain may accomplish VOR adaptation (Lisberger and Sejnowski, 1991). "Now I would be the first to say," Sejnowski emphasized, "that this is a very simple model, but it has already taught us a few important lessons about how different behaviors can interact and how changing a system as complex as the brain (for example by making a lesion) can have effects that are not intuitive."

Sejnowski has been very active in the connectionist movement and has been credited with "bringing computer scientists together with neurobiologists" (Barinaga, 1990). With Geoffrey Hinton he created in 1983 one of the first learning networks with interneurons, the so-called hidden layers (Hinton and Sejnowski, 1983). Back propagation is another learning scheme for hidden units that was first introduced in 1986 (Rumelhart et al., 1986), but its use in Sejnowski's 1987 NETtalk system suddenly dramatized its value (Sejnowski and Rosenberg, 1987). NETtalk converts written text to spoken speech, and learns to do so with increasing skill each time it is allowed to practice. Said Koch: "NETtalk was really very very neat. And it galvanized everyone. The phenomena involved supervised learning." The network, in trying to interpret the pronunciation of a phoneme, looks at each letter in the context of the three letters or spaces on either side of it, and then it randomly activates one of its output nodes, corresponding to the pronunciation of a certain phoneme. "The supervised learning scheme now tells the network whether the output was good or bad, and some of the internal weights or synapses are adjusted accordingly," explained Koch. "As the system goes through this corpus of material over and over again—learning by rote—you can see that through feedback and supervised learning, the network really does improve its performance while it first reads unintelligible gibberish, at the end of the training period it can read text in an understandable manner, even if the network has never seen the text before. Along the way, and this is fascinating, it makes a lot of the kinds of mistakes that real human children make in similar situations."

THE LEAP TO PERCEPTION

At the Frontiers symposium, Sejnowski also discussed the results of an experiment by Newsome and co-workers (Salzman et al., 1990), which Sejnowski said he believes is "going to have a great impact on the future of how we understand processing in the cortex." Working with monkeys trained to indicate which direction they perceived a stimulus to be moving, Newsome and his group microstimulated the cortex with an electrical current, causing the monkeys to respond to the external influences on their motion detectors, rather than to what they were "really seeing." The results, said Sejnowski, were "very highly statistically significant. It is one of the first cases I know where with a relatively small outside interference to a few neurons in a perceptual system you can directly bias the response of the whole animal."

Another experiment by Newsome and his colleagues (Newsome et al., 1989) further supports the premise that a certain few motion-detecting cells, when triggered, correlate almost indistinguishably with the animal's overall perception. This result prompted Koch to ask how, for instance, the animal selects among all MT neurons the proper subset of cells, i.e., only those neurons corresponding to the object out there that the animal is currently paying attention to." The answer, he thinks, could lie in the insights to be gained from more experiments like Newsome's that would link the reactions of single cells to perception.

Considering his own, these, and other experiments, Koch as a neuroscientist sees "something about the way the brain processes information in general," he believes, not just motion-related phenomena. To meet standards of reasonableness, models concerned with biological information processing must take account of the constraints he discovered crucial to his own model, Koch thinks. First, the algorithm should be computationally simple, said Koch: "Elaborate schemes requiring very accurate components, higher-order derivatives," and the like are unlikely to be implemented in nervous tissue. The brain also seems to like unit/place and population coding, as this approach is robust and comparatively impervious to noise from the components. Also, algorithms need to remain within the time constraints of the system in question: "Since perception can occur within 100 to 200 milliseconds, the appropriate computer algorithm must find a solution equally fast," he said. Finally, individual neurons are "computationally more powerful than the linear-threshold units favored by standard neural network theories," Koch explained, and this power should be manifest in the algorithm.

These insights and many more discussed at the symposium reflect the least controversial of what neural networks have to offer, in that they reflect the determination, where possible, to remain valid within the recognized biological constraints of nervous systems. It can be readily seen, however, that they emanate from an experimental realm that is comparatively new to scientific inquiry. Typically, a scientist knows what he or she is after and makes very calculated guesses about how to tease it into a demonstrable, repeatable, verifiable, physical—in short, experimental—form. This classical method of inquiry has served science well, and will continue to do so. But the brain—and how it produces the quality of mind—presents to modern science what most feel is its greatest challenge, and may not yield as readily to such a straightforward approach. While more has been learned about the brain in the last three decades of neuroscience than all that went before, there is still a vast terra incognita within our heads. People from many disciplines are thinking hard about these questions, employing an armamentarium of scientific instruments and probes to do so. But the commander of this campaign to unravel the mysteries of the human brain is the very object of that quest. The paradox inherent in this situation has been a subject of much concern, as suggested in a definition proposed by Ambrose Bierce (Allman, 1989, p. 5):

> MIND, *n.*—A mysterious form of matter secreted by the brain. Its chief activity consists in the endeavor to ascertain its own nature, the futility of the attempt being due to the fact that it has nothing but itself to know itself with.

At last there may be more than irony with which to confront the paradox. Neural networks do learn. Some would argue that they also think. To say so boldly and categorically embroils one in a polemic, which—considering the awesome implications of the proposition—is perhaps as it should be. But whatever the outcome of that debate, the neural network revolution seems on solid ground. It has already begun to spawn early, admittedly crude, models that in demonstration sometimes remind one of the 2-year-old's proud parents who boasted about how well their progeny was eating, even as the little creature was splashing food all over itself and everyone nearby. And yet, that same rough, unformed collection of human skills will grow—using algorithms of learning and training, some inherent, others largely imposed by society—into what could become an astounding intellect, as its brain uses the successive years to form, to learn, and to work its magic. Give the neural networkers the same time frame, and it seems altogether possible that their network creations will grow with comparable—from here nearly unimaginable—power and creativity. Wrote Koch and Segev, "Although it is difficult to predict how successful these brain models will be, we certainly appear to be in a golden age, on the threshold of understanding our own intelligence" (Koch and Segev, 1989, p. 8).

BIBLIOGRAPHY

Adelson, E.H., and J. Movshon. 1982. Phenomenal coherence of moving visual patterns. Nature 200:523-525.

Allman, William F. 1989. Apprentices of Wonder: Inside the Neural Network Revolution. Bantam Books, New York.

Barinaga, Marcia. 1990. Neuroscience models the brain. Science 247:524-526.

Bloom, Floyd E., and Arlyne Lazerson. 1988. Brain, Mind, and Behavior. Freeman, New York.

Changeux, Jean-Pierre. 1986. Neuronal Man. Oxford University Press, New York.

Churchland, Patricia S., Christof Koch, and Terrence J. Sejnowski. 1990. What is computational neuroscience? Pp. 46-55 in Computational Neuroscience. Eric L. Schwartz (ed.). MIT Press, Cambridge, Mass.

Churchland, Paul M., and Patricia Smith Churchland. 1990. Could a machine think? Scientific American 262(January):25-37.

Crick, Francis, and Christof Koch. 1990. Towards a neurobiological theory of consciousness. Seminars in the Neurosciences 2:263-275.

Harris, J., C. Koch, E. Staats, and J. Luo. 1990. Analog hardware for detecting discontinuities in early vision. International Journal of Computer Vision 4:211-223.

Hinton, G.E., and T.J. Sejnowski. 1983. Optimal perceptual inference. Proceedings of the IEEE Computer Society Conference on Computer Vision and Pattern Recognition. Washington, D.C., June.

Hinton, Geoffrey, and J.A. Anderson. 1981. Parallel Models of Associative Memory. Lawrence Ellbaum, Assoc., Hillsdale, N.J.

Hopfield, J.J. 1982. Neural networks and physical systems with emergent collective computational abilities. Proceedings of the National Academy of Sciences of the U.S.A. 79:2554-2558.

Hutchinson, J., C. Koch, J. Ma, and C. Mead. 1988. Computing motion using analog and binary sensitive networks. IEEE Computer 21:52-63.

Jacob, Francois. 1982. The Possible and the Actual. University of Washington Press, Seattle.

Koch, C., T.H. Wang, and B. Mathur. 1989. Computing motion in the primate's visual system. Journal of Experimental Biology 146:115-139.

Koch, Christof, and Idan Segev (eds.). 1989. Methods in Neuronal Modeling. MIT Press, Cambridge, Mass.

Lisberger, S.G., and T.J. Sejnowski. 1991. Computational analysis suggests a new hypothesis for motor learning in the vestibulo-ocular reflex. Technical Report INC-9201. Institute for Neural Computation, University of California, San Diego.

Newsome, W.T., K.H. Britten, and J.A. Movshon. 1989. Neuronal correlates of a perceptual decision. Nature 341:52-54.

Pagels, Heinz R. 1988. The Dreams of Reason: The Computer and the Rise of the Sciences of Complexity. Simon and Schuster, New York.

Rumelhart, D., G. Hinton, and R. Williams. 1986. Learning internal representations by error propagation. Pp. 318-362 in Parallel Distributed Processing, Volume I. D. Rumelhart and J. McClelland (eds.). MIT Press, Cambridge, Mass.

Rumelhart, D., J. McClelland, and the PDP Research Group. 1986. Parallel

Distributed Processing: Explorations in the Microstructure of Cognition. MIT Press, Cambridge, Mass.

Salzman, C.D., K.H. Britten, and W.T. Newsome. 1990. Cortical microstimulation influences perceptual judgments of motion direction. Nature 346:174-177.

Searle, John R. 1990. Is the brain's mind a computer program? Scientific American 262(January):25-37.

Sejnowski, T.J., and C.R. Rosenberg. 1987. Parallel networks that learn to pronounce English text. Complex Systems 1:145-168.

Sejnowski, T.J., C. Koch, and P.S. Churchland. 1988. Computational neuroscience. Science 241:1299-1306.

Sherrington, C. The Integrative Action of the Nervous System [first published in 1906]. Reprinted by Yale University Press, New Haven, 1947.

Weiner, N. 1947. Cybernetics. MIT Press, Cambridge, Mass.

RECOMMENDED READING

Churchland, P.S. 1986. Neurophilosophy: Toward a Unified Science of the Mind/Brain. MIT Press, Cambridge, Mass.

Churchland, P.S., and Sejnowski, T.J. 1992. The Computational Brain. MIT Press, Cambridge, Mass.

Durbin, R., C. Miall, and G. Mitchison (eds.). The Computing Neuron. Addison-Wesley, Reading, Mass.

Hildreth, E.C. 1984. The Measurement of Visual Motion. MIT Press, Cambridge, Mass.

Hubel, D.H. 1988. Eye, Brain and Vision. Scientific American Library, New York.

Ullman, S. 1981. Analysis of visual motion by biological and computer systems. IEEE Computer 14:57-69.

10

Quasicrystals and Superconductors: Advances in Condensed Matter Physics

by EDWARD EDELSON

The physics session of the 1990 Frontiers of Science symposium was devoted to two separate and unusual topics in condensed matter physics (the physics of solid and liquid objects), which the chairman, Daniel Fisher of Harvard University, described as including everything "that is much bigger than a molecule, much smaller than a mountain, and not too hot"—a range diverse enough to occupy almost half of all physicists. Both topics—high-temperature superconductivity and quasicrystals—concerned fields that "seemed to be, if not dead, at least not so active in the 1970s and the early 1980s" but suddenly came to life as the result of major, surprising discoveries, Fisher said.

High-temperature superconductivity, which has received wide coverage in the popular press, has become the focus of intensive effort among applied and theoretical physicists and materials scientists. The discovery in 1986 by J. Georg Bednorz and Karl Alex Mueller of the IBM Research Laboratories in Zurich, Switzerland, of a new class of materials that lose all electrical resistance at unprecedentedly high temperatures has captured professional interest as well as public attention. Visions of levitated trains, new energy sources, and ultra-accurate scientific instruments have energized a field in which incremental advances had been the norm for decades. The discovery also transformed superconductivity from "the most well-understood problem in solid-state physics," in the words of Alex Zettl of the University of California, Berkeley, to a field where a thousand theoretical flowers have bloomed in efforts to explain why the newly discovered materials are capable of high-temperature superconductivity.

The other development discussed in the physics session, although less publicized in the lay press, has had an equally revolutionary effect on the field of crystallography. It is the discovery of a class of materials that violate the rigorous, long-established rules about crystals—solids that consist of regular, repeating, three-dimensional units. Theory had held that certain structures were forbidden by nature. Now structures have been found to exist that are neither glasses nor crystals. They are not composed of the repeating, three-dimensional units of crystals, and they exhibit symmetries found neither in crystals nor in glasses. These quasicrystals, as they are called, are a fundamentally new, ordered state of matter.

The unusual nature of a quasicrystal is best explained by a two-dimensional analogy, the tiling of a floor or other surface. We customarily cover a floor with square tiles, which can be said to have fourfold symmetry because they have four equal sides. A surface can also be covered with triangular tiles, which have threefold symmetry, and with tiles that have sixfold symmetry. But it cannot be covered completely by pentagons, which have fivefold symmetry. No matter how cleverly we lay pentagonal tiles, gaps are left that cannot be filled using those tiles. In the same way, a three-dimensional space can be filled periodically with crystal units that have fourfold or sixfold symmetry but not, according to theory, by crystal units with fivefold symmetry. That theory now has been upset by the discovery of crystal units that have fivefold symmetry and fill space completely. As in the case of superconductivity, this discovery has excited the interest of physicists, who are studying the properties of quasicrystals and how they are made in nature, as well as of theorists, who are exploring the mathematical and physical implications of the existence of quasicrystals.

SUPERCONDUCTIVITY

A Brief History

The phenomenon of superconductivity was discovered in 1911 by a Dutch physicist, H. Kamerlingh Onnes, who found that the electrical resistance of mercury vanished suddenly when the metal was cooled to a temperature of about 4 kelvin (K), which is 4 degrees Celsius above absolute zero (Table 10.1). If an electrical current is established in a ring of frozen mercury that is maintained at that temperature, the current will persist indefinitely. By contrast, such a current dies away quickly in an ordinary conducting material such as copper.

In the 1930s, another characteristic of superconducting materials was described. If an ordinary metal is placed in a magnetic field, the magnetic field permeates the material. Superconducting materials act differently. Some of them expel the magnetic field completely; others allow only partial penetration. This Meissner effect, as it is called, is responsible for the ability of permanent magnets to be levitated above a superconductor. In expelling the field of the magnet, the superconductor generates its own magnetic field, which pushes the magnet away and allows the magnet to float over the superconducting sample.

A full microscopic theory of superconductivity was achieved in the 1950s by three physicists, John Bardeen, Leon N. Cooper, and J. Robert Schrieffer. The Bardeen-Cooper-Schrieffer (BCS) theory starts with the picture of a normally conducting metal whose atoms are arranged in a three-dimensional crystal structure. Some of the loose-

TABLE 10.1 Developments in Superconductivity

Theory	Date	Experiment
No theory	1911	Superconductivity discovered $T_c = 4.2$ K
London equations	1930	Meissner effect
BCS theory	1950	Isotope effect
Superconductivity: "Most well-understood problem in solid-	1960	Type II materials, Josephson effects
state physics"	1970	$T_c = 23$ K
High-T_c theory?	1986	$T_c = 30$ K
	1987	$T_c = 90$ K
	1988	$T_c = 125$ K Copper oxides

SOURCE: Courtesy of A. Zettl.

ly held outer electrons drift away from the metal atoms in the crystal, forming an electron gas that flows freely through the crystal lattice. The flow of these electrons is the electrical current that runs through the metal. These electrons do not have limitless freedom, however. Some of them interact with impurities in the metal and with the vibrations of the atoms that form the crystal lattice. The interaction of the electrons with the atomic vibrations causes the electrical resistance found in metals at ordinary temperatures.

Electrons, which have a negative charge, ordinarily repel each other. The essence of the BCS theory is that under some circumstances they can have a net attractive interaction and thus form pairs. The BCS theory pictures an electron traveling through a lattice of metal ions, which are positively charged because they have lost some of their electrons. The negatively charged electron may be attracted to another negatively charged electron by way of the oscillations of the metal ions. The two electrons form a pair—a Cooper pair, in the language of the BCS theory. All the Cooper pairs act together as a unified quantum system. Circulating currents made up of Cooper pairs do not decay in the ordinary manner of currents composed of single electrons.

One useful way to picture the interaction is to say that one negatively charged electron attracts the positively charged ions around it, causing a slight distortion in the lattice. The distortion produces a polarization, an area of increased positive charge. A second electron is attracted by this pocket of positive charge; it thus becomes coupled to the first electron, following it through the lattice.

Another description of the same phenomenon is that the electrons are coupled by the interchange of a virtual particle, the phonon. Phonons represent the vibrations of the lattice. In either picture, the electron–phonon coupling allows the electrons to pair and then to flow unhindered through the lattice. Anything that destroys the Cooper pairs—for example, heat that increases the lattice vibrations above a certain limit—destroys superconductivity.

Among other things, the BCS theory explains why superconductivity occurs in metals only at very low temperatures. At higher temperatures, thermal motion begins to break Cooper pairs apart. Above a given transition temperature, all the pairs are broken and superconductivity vanishes. BCS theory also explains why metals that are good conductors at room temperature may not be superconductors at low temperatures: they do not have a large enough phonon–electron interaction to allow Cooper pairs to form.

In the BCS theory, a material's transition temperature, the temperature at which it becomes superconducting, depends only on three factors: the phonon frequency in the lattice, the density of states (energy levels that electrons can occupy) near the Fermi energy (the highest electron energy occupied in the material at low temperature), and the strength of the electron–phonon coupling energy.

Another feature of superconducting materials is the isotope effect. If an atom of one element in a superconducting material is replaced by an isotope of greater mass, the transition temperature of the material generally goes down. This so-called isotope effect occurs because transition temperature is approximately proportional to the frequency of the lattice vibrations, and isotopes of greater mass have lower vibration frequencies. The isotope effect, first demonstrated in mercury in 1950, was fundamental to the development of the BCS theory, because it strongly implied that phonons were the glue that held Cooper pairs together.

The success of the BCS theory came against a background of slow, dogged advances in superconducting materials research. Starting from the 4.15-K transition temperature of mercury described by Onnes in 1911, physicists discovered a series of materials with progressively higher transition temperatures. By the mid-1970s, the record transition temperature was 23 K, in a niobium-germanium compound.

For practical purposes, characteristics other than transition temperature are also important. Superconductivity can be destroyed by a high current density or a strong magnetic field. Technologically, the most desirable superconducting material has a high transition temperature and a high critical current and remains superconducting in a strong magnetic field. One such material that combines these qualities is a niobium-titanium compound that is used in superconducting magnets such as those found in medical magnetic resonance imaging devices and those being built for the Superconducting Super Collider, the world's largest particle accelerator.

Another property of superconductors, described by Brian Josephson in the early 1960s, is the Josephson effect. If two samples of superconducting material are separated by a thin barrier, some of the Cooper-paired electrons will tunnel through. This tunneling is explained by quantum theory, in which subatomic particles are described equally well as wave packets. It is the wave aspect of electrons that allows tunneling to occur. The electron as a particle cannot pass through a barrier; as a wave, it can. The Josephson effect has led to the development of high-accuracy superconducting electronic instruments, such as ultrasensitive magnetometers, that use the tunneling effect.

By the late 1980s, therefore, superconductivity was a well-understood if hardly dynamic field. BCS theory explained the phenomenon but provided no help in finding more useful superconductors. Magnet makers had mastered the skill of working with the available superconductors but were chafing under the necessity of cooling the materials to a few degrees above absolute zero. Those temperatures could be achieved only by the use of liquid helium as the coolant in extremely expensive refrigeration systems. The quest taken up in superconductivity research was to find a material that would be superconducting at a temperature above 77 K, which would allow use of inexpensive refrigeration systems with nitrogen (which becomes liquid at 77 K) as the coolant. But such an advance was nowhere in sight. The highest known transition temperature had increased by an average of 0.3 K per year for several decades. Extrapolation of a line through points on a graph representing new highs in transition temperature indicated that liquid nitrogen temperatures would be reached late in the 21st century. Occasionally, one laboratory or another would make a preliminary report of a material with extremely high transition temperatures, but none of those reports stood up under scrutiny.

Start of a New Era

Thus there was great excitement when Bednorz and Mueller reported in 1986 an apparent transition temperature of 30 K in a material completely different from the known superconductors, a ceramic made up of copper, oxygen, barium, and lanthanum. This time the claim was verified, and the world of superconductivity entered a new era. Within a matter of months, Paul Chu at the University of Houston and Maw-Kuen Wu at the University of Alabama reported that a copper oxide ceramic containing yttrium and barium had a transition temperature well above 90 K, easily in liquid nitrogen territory. The record now is a transition temperature of 125 K, in a copper oxide that contains thallium, barium, and calcium (Figure 10.1).

Challenges Posed by High-temperature Superconducters

These high-temperature superconductors pose a number of challenges. One is the practical matter of making devices from these materials. Since the new materials are ceramics, they are not easily made into wires for magnets. In addition, although their transition temperatures are high, their critical currents are not; they can carry only a small fraction of the current that can flow through an ordinary copper wire without losing their superconductivity. A substantial

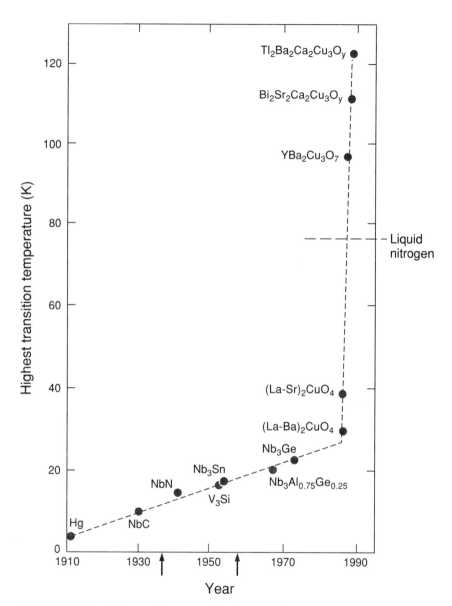

FIGURE 10.1 History of progress in the quest for an ever-higher transition temperature. (Courtesy of A. Zettl.)

effort in academic and industrial laboratories has started to solve some of these problems.

Physicists have also been challenged to provide a theory for high-temperature superconductivity. The atomic structure of the new superconductors is quite different from that of the known materials. The outstanding feature of the new high-temperature materials is that they are oxides, most of which are electrical insulators. They have planes consisting of atoms of copper and oxygen, with two copper atoms for each oxygen. The other atoms, such as barium and yttrium, are between the copper-oxygen planes. In the language of physics, these materials are highly anisotropic: the two-dimensional planes of atoms dominate the structure. One assumption of many physicists was that the mechanism of superconductivity for these new materials might be closely related to this structural anisotropy.

The implication was that high-temperature superconductivity might well not be explained by the BCS theory—an implication strengthened by the belief that the theory could not accommodate such high transition temperatures. Investigations to see whether BCS theory or some variation on it could explain the new superconductivity began, and are still continuing.

A first approach to the issue was to determine whether the electrons in the new materials formed pairs during the superconducting phase. As Zettl explained, there are two standard ways to make this determination. One is to make a ring of the material. The magnetic flux through the ring is quantized, and measurements of the quantum of flux can determine whether the unit of charge that produces the magnetic flux is the charge on one electron, e, or two electrons, 2e. If the charge is e, the electrons are not paired; if it is 2e, they are.

A second method is to make a Josephson junction, allow electron tunneling to occur, and shine microwaves of a known frequency on the junction. The interference between this frequency and an internal frequency generated by the tunneling electrons indicates whether the charge is e or 2e.

Both experiments were done within months, and both indicated an effective charge of 2e: electrons in high-temperature superconductors are paired. "It is pretty clear that we have a pairing mechanism, and the question then becomes what mechanism is holding these pairs together," Zettl said.

The BCS theory makes some specific predictions about the properties of electron pairs. Calculations done by Marvin Cohen and co-workers indicated that the BCS theory could accommodate a transition temperature as high as 100 K.

However, experiments done by Zettl and others to measure the isotope effect in the new materials produced results that complicated the issue. In these experiments, the oxygen-16 atoms in the material were replaced by oxygen-18 isotopes. The predicted result given by the BCS theory was a 5-K reduction in the transition temperature. The actual reduction was a small fraction of 1 K. Further calculations showed that the BCS theory could be consistent with a small or non-existent isotope effect under certain circumstances. And other calculations showed that the quasi-two-dimensional system of the new materials could also suppress the isotope effect. But it is almost impossible to accommodate both the observed high transition temperatures and the small isotope effect within BCS theory, Zettl said— one of the many unresolved issues about the new materials.

The issue of dimensionality has become very important. Several experimental programs have examined whether the superconducting current flows only along the planes of the materials or if it is three-dimensional. Some experiments tried to determine whether the new materials behaved as a collection of weakly coupled slabs, with all the action associated with superconductivity limited to the copper-oxygen slabs. In one group of experiments designed to test this idea, layers of nonsuperconducting copper oxide were inserted between the copper-oxygen layers of a superconducting compound. The material remained a superconductor even when the insulating layers became very thick, but the transition temperature of the superconducting material went down. In another experiment, Zettl and his collaborators intercalated iodine and other atomic species between the copper-oxygen layers, steadily increasing the distance between the layers of a superconducting ceramic. Even with a 23 percent expansion of the vertical distance of the material, the transition temperature scarcely changed. "This is evidence that the superconductivity does not strongly rely on three-dimensional coupling between superconducting sheets, with obvious implications for two-dimensional mechanisms," Zettl said.

Another question concerned what is called the energy gap in these materials. The energy gap consists of a range of energies that electrons in the material cannot achieve; measurement of the energy gap can give the binding energy and coherence length (the distance between electrons) of the electron pairs. The existence of a gap, and its magnitude if it does exist, are important theoretically. One class of proposed theories about the new materials is based on high binding energy for the electron pairs, the so-called strong-coupling limit, while another class is based on low binding energy, the so-called weak-coupling limit.

One measurement used to describe a superconducting material is a dimensionless number derived by multiplying the energy gap by 2 and dividing it by the Boltzmann constant times the transition temperature. In the classic BCS theory, in the weak-coupling limit, the value of this expression is about 3.5. A number larger than 3.5 indicates strong electron coupling and hence a large gap.

Values for this number can be derived in several ways. One is to use nuclear magnetic resonance (NMR) to examine a material. A sample is placed in a magnetic field and radio waves of appropriate frequency are beamed in. The relaxation time of nuclei as they give up the absorbed energy gives information about the properties of the sample. NMR data give values for the yttrium-barium-copper oxide in the range of 8 to 11. Another method is to study the absorption of infrared light by the material. Using this method of infrared spectroscopy, experimenters have obtained values between 3.5 and 8.

A third method is tunneling, in which a voltage is applied to a metal tip in contact with the superconducting material. If the voltage is sufficient to break the Cooper pairs, electrons will tunnel through. The applied voltage is thus a measure of the superconducting gap. "From such tunneling measurements on the superconducting oxide bismuth-strontium-calcium-copper-oxygen, two distinct values for the energy gap have been determined: 3.5 and 7," Zettl said. "The two values refer to different crystallographic orientations. The smaller value is consistent with BCS theory, while the larger value is not."

The issue of strong binding versus weak binding of the electron pairs thus remains confusing, said Duncan Haldane of Princeton University, another speaker at the symposium, who focused on the theoretical issues.

One critical factor in trying to develop a theory to explain the new materials, Haldane said, is their odd behavior in terms of electrical resistance above the transition temperature. When a metal such as copper is cooled, its resistance decreases linearly at high temperatures and then undergoes a very rapid decrease at lower temperatures. In the new materials, the resistance decreases linearly to much lower temperatures than expected, disappearing suddenly at the transition temperature. "This linear resistance above the superconducting transition is difficult to explain within the usual pictures of conductivity in metals," Haldane said.

Development of New Theories

The new theories can be grouped into two classes. One group assumes weak coupling, whereas a second assumes strong coupling. Most of the theories assume that polarization causes electron pairing.

In the electron-phonon interaction of the BCS theory, the electron polarizes the lattice. Two other things can be polarized, charge or spin. A huge variety of exotic polarizing mechanisms—excitons, plasmons, spin fluctuations—have been proposed.

A theory based on charge polarization postulates the excitonic mechanism, first outlined by Vitaly Ginzburg of the Soviet Union and independently by William Little of Stanford University and later developed by Bardeen, David Allender, and James Bray at the University of Illinois. It pictures electrons as being paired by exchange of a virtual particle, the exciton, that comes into existence when an electron moves to a higher energy state, creating a hole, and then drops back to the lower energy state. It can be put in another way: the electron creates a polarization cloud by pushing nearby electrons away because of the Coulomb repulsion between similarly charged particles. A second electron has its energy lowered because of the polarization cloud and forms a pair with the first. One problem with the excitonic mechanism is that it predicts transition temperatures that are too high—by hundreds or even thousands of kelvins. In addition, the requisite excitations have not been observed.

A spin-polarization theory first developed by P.-G. de Gennes assumes the production of virtual particles called magnons. It starts with a nonsuperconducting material in the antiferromagnetic state, in which neighboring electrons have spins of opposite orientation. When the material is doped by the addition of another element, holes are created that also carry a spin. Each hole wants to spin antiparallel to the electrons, but it also wants to delocalize—spread its wave function to other sites. But it sees antiferromagnetic electrons whose spins are aligned parallel with it at these other sites. Its spin is tilted, so that the electron can delocalize to some degree and still have a favorable alignment of its spin with the antiferromagnetic electrons. This causes a spin polarization cloud to form, attracting a second electron and forming a pair. The interaction is mediated by polarizing spin degrees of freedom.

The charge-polarization and spin-polarization theories are based on two sharply different pictures of the electron structure of the material. BCS theory assumes that the conduction electrons that form a Cooper pair are delocalized and move relatively independently. Charge-polarization theories also are based on this structure. Many spin-polarization theories assume a strikingly different structure, called the Mott-Hubbard insulator.

As described by Sir Nevill Mott, a British Nobel laureate in physics, and John Hubbard, a Briton who was with IBM before his death, the Mott-Hubbard insulator picture assumes that the electrons are in an ordered, strongly correlated structure. The atoms in the lattice have outer electron shells that are almost exactly half-filled, so that there is one conduction electron per atom and the material is an insulator. Doping the material produces mobile holes. The collective behavior responsible for the transition to superconductivity occurs through correlations of the electron spins.

Another spin-polarization theory based on the Hubbard model pictures strong coupling of the electrons based on antiferromagnetic spin fluctuations. The charge carriers are holes—missing electrons— that are created as barium is added to lanthanum copper oxide. Cooper pairing occurs as a result of interactions between the spins of the holes and neighboring electrons.

One theory that aroused enthusiasm for a while was based on exotic particles called anyons or semions, which are neither bosons nor fermions, Haldane said. A pair of fermions can form a boson; so can a pair of bosons, he explained. In a two-dimensional system, there is a third possibility: Two semions can pair to form a boson, too. The anyon theory is that the boson sea of a high-temperature superconductor is formed by as-yet-undetected particles whose properties are halfway between those of fermions and bosons. The anyon theory requires time-reversal breaking—a violation of the rule that a movie made of a particle interaction would show the exact reverse of that interaction when run backward. Experiments done at Bell Laboratories in 1990 indicated that time reversal was violated in the new superconductors, a finding that gave prominence to the anyon theory. Interest has faded as other laboratories have failed to replicate the broken time-reversal studies, Haldane said.

Yet another theory assumes the existence of "soft" phonons associated with what is called a breathing mode, in which oxygen atoms that are around copper ions in the lattice move in and out regularly. As temperature drops, the breathing mode should soften—that is, oscillate at lower and lower frequencies. Below a given temperature, the oscillation stops. The resulting "frozen phonons" allow strong electron coupling that permits high transition temperatures.

Another theory posits interactions mediated by bipolarons. A polaron is an electron that sits in a pocket of positive charge in a lattice; a bipolaron is a pair of such electrons. Bipolarons can undergo a Bose condensation if the electron structure of a material meets certain conditions that can occur in layered materials such as the oxide superconductors.

Philip W. Anderson of Princeton University has proposed a resonance valence bond (RVB) theory that assumes no polarization of any kind. It postulates a spin liquid, in which electrons can be described as being in a kind of square dance, with overall order maintained even as dancers constantly change partners. RVB theory holds that electron pairs can exist in material that is not superconducting. Superconductivity occurs when the material is doped to create holes, which form pairs and undergo Bose condensation.

None of the theories has won general acceptance, Haldane said, nor is any resolution of the issue in sight.

Progress Toward Practical Applications

But just as was true of the low-temperature superconductors, progress toward practical applications of the new materials is being made, said Zettl and Peter Gammel of Bell Laboratories, another speaker at the Frontiers symposium. Both spoke about the issue of preserving superconductivity in strong magnetic fields, which is essential for making superconducting magnets.

The copper oxide superconductors, like the low-temperature superconductors now used in large magnets, do not expel magnetic fields completely. Instead, the magnetic field penetrates the material in some areas, forming vortices that are not superconducting; the rest of the material remains superconducting. At sufficiently high magnetic fields, the vortices can pervade the sample, and superconductivity vanishes. Current flowing through the material can cause the vortices to move, which likewise destroys superconductivity. The critical current is that which causes the vortices to move enough to destroy the zero-resistance state.

A new theory and its experimental verification have provided an explanation for this behavior of the high-temperature superconductors. In low-temperature superconductors, the problem has been solved by introducing defects into the material. The defects "pin" the magnetic field lines, thus preventing them from interfering with superconductivity.

The anisotropy of the new superconductors makes this method much less effective. Their layered structure means that a magnetic line that is pinned by a defect in one layer may not be pinned by a corresponding defect in other layers. As a result, the transition to superconductivity in these materials does not occur in a narrow range of temperatures. Instead, resistivity tends to decrease gradually; in some cases, the material never loses all resistance in a strong magnetic field. This phenomenon appeared to present a major barrier to use of the high-temperature superconductors in magnets, one of the major applications of superconducting materials.

A theory proposed by Matthew Fisher of IBM offered a way out of this dilemma, at least for yttrium-based superconductors. Fisher started with the standard picture of magnetic field lines going through the superconductor in bundles, or vortices. It is these vortices that are pinned by defects in low-temperature superconductors. Fisher proposed that in the new material, interactions between the vortices can cause them to freeze in place at low temperatures, in a sudden transition that allows the material to be a true superconductor. According to this theory, the phenomenon is analogous to the freezing of water, which is liquid until the temperature is lowered sufficiently. A high-temperature superconductor thus would be a "vortex glass," because the vortices would be frozen at a transition temperature.

Others elaborated on the theory, producing precise predictions that could be tested experimentally. One such prediction was that the superconductor would undergo a sharp phase transition, losing all resistivity, at a given temperature; it would be the transition to the vortex glass state. Other predictions about the transition to the vortex state could be made, but they were difficult to measure because experiments required a degree of precision that was at the boundary of existing instrumentation.

Recently, Gammel and others have done confirmatory experiments on yttrium-containing superconductors. The results were in agreement with the predictions of the vortex glass theory. Experiments on other high-temperature superconductors are ongoing.

Verification of the vortex glass theory has important implications for physicists working on industrial uses of the new superconductors. For example, the theory might help researchers introduce defects into superconducting materials in a way designed to make the transition to the vortex glass state, and thus to superconductivity, occur at higher temperatures. In this case, fundamental theory has had a direct impact on practical applications of a promising new technology.

The greatest progress has been made in the field of miniature electronic instruments using thin films of high-temperature superconductors, Zettl said: "Our electronics industry is based on thin-film techniques, and these materials lend themselves rather well to thin-film fabrication."

QUASICRYSTALS

Solid materials have traditionally been divided into two broad classes, based on atomic order. One class consists of the glasses. These are solids whose atoms are in a random, close-packed arrangement. A glass can be regarded as a liquid whose flow is immeasurably slow.

The second class consists of the crystals, whose atoms are ordered in a regular, periodically repeating three-dimensional pattern. Crystals can be decomposed into clusters of atoms called unit cells. A crystal consists of a large number of close-packed, identical unit cells—in theory, an infinite number—which extend in all directions and fill space completely. One property of a crystal is that all the unit cells have the same orientation in space. Consequently, a crystal can overlie itself only when it is rotated by specific angles through specific points. A unit cell that has twofold symmetry must be rotated by 180 or 360 degrees to overlie itself; if it has threefold symmetry, it must be rotated by 120, 240, and 360 degrees. Only twofold, threefold, fourfold, and sixfold rotational symmetries are possible in crystals, because only unit cells with these symmetries can fill three-dimensional space completely. Fivefold symmetry is strictly forbidden.

As mentioned previously, there is a close analogy between crystals and two-dimensional tilings. A surface can be covered completely by rectangles, which have twofold symmetry; by square tiles, which have fourfold symmetry; by triangular tiles, which have threefold symmetry; or by hexagonal tiles, which have sixfold symmetry. It cannot be covered completely by pentagonal tiles, which have fivefold symmetry.

But physicists now have discovered a class of solid materials that are neither glasses nor crystals; they have been named quasicrystals. Peter Bancel, a physicist then at the IBM Watson Research Center in Yorktown Heights, N.Y. (and now at the Center for the Studies of Chemical Metallurgy in Paris), who spoke at the symposium, described quasicrystals as "ordered atomic solids that possess long-range quasiperiodic positional order and a long-range non-crystallographic orientational order." In other words, they fill space completely without having the symmetry and order of classic crystals. Their atomic arrangements exhibit overall symmetries, such as fivefold, eightfold and tenfold symmetries, that have never before been observed. "The breakthrough raises the possibility of a whole class of materials with surprising electronic, vibrational, and thermal properties," said Paul J. Steinhardt of the University of Pennsylvania, a participant in the symposium and a leading figure in the story of quasicrystals.

The first quasicrystal was discovered in 1982 by Dany Shechtman, an Israeli physicist working at the National Bureau of Standards (now the National Institute of Standards and Technology) outside Washington, D.C. Schechtman was studying the properties of alloys formed by very rapid cooling. When he quenched a liquid mixture of aluminum and a small amount of manganese, he produced a solid composed of micron-sized grains with feathery arms. The unusual nature of the material became apparent when it was viewed by electron diffraction, a procedure in which a beam of electrons is directed through a solid. The patterns resulting from the scattered, or diffracted, electrons in the material can be interpreted to indicate the atomic structure and symmetry of the solid.

The diffraction pattern of the aluminum-manganese alloy consisted of sharp spots arranged with fivefold symmetry; coincidence could thus be achieved by rotations of 72 degrees. The diffraction pattern indicated that the material consisted of a space-filling atomic structure with the symmetry of an icosahedron. (An icosahedron is a regular polyhedron with 20 identical triangular faces and six fivefold symmetry axes.) The observation of this icosahedrally symmetric pattern was startling precisely because an icosahedron, rotated about one of its axes, has the fivefold rotational symmetry that conventional crystallography held to be impossible for a crystal. At about the time that Shechtman made his discovery, Paul Steinhardt and his graduate student, Dov Levine, had been trying to understand the properties of glassy solids. Earlier, Steinhardt had been running computer simulations of rapid cooling of a liquid made of idealized spherical atoms. The readout showed a glass that included some clusters of symmetrical, icosahedral atomic structures. Steinhardt and Levine then proposed the theoretical possibility that quasicrystals with fivefold symmetry could exist. As the story developed, this theoretical model incorporated many of the features actually seen in the new alloys, and it has emerged as the leading explanation for them.

Steinhardt had started from a consideration of two-dimensional nonperiodic tilings. It was then known that a surface can be covered by a tiling with two or more tile types in which the tile order is quasiperiodic: the tiles repeat in a sequence made up of subsequences whose periods are in an irrational ratio. (An irrational number is one that cannot be expressed as the ratio of two integers. Pi is an irrational number, as is the square root of 2.) The overall sequence never quite repeats perfectly, but the surface is nonetheless covered completely.

The study of these nonperiodic two-dimensional tilings had been going on since the 1960s. In 1966, Robert Berger of Harvard University showed that a surface could be covered completely by a nonperiodic tiling. Berger described two solutions, one with more than 20,000 types of tile and one with 104 types.

The British mathematical physicist Roger Penrose of Oxford University ultimately found a solution with just two tile types. He showed how two types of four-sided tiles, both rhombi (a rhombus is a parallelogram with all sides of equal length), a fat and thin diamond, could cover a surface nonperiodically. The two fundamental shapes in a "Penrose tiling" are arranged in a pattern with fivefold orientational order.

Steinhardt and Levine extended Penrose's tiling to three dimensions, working out a way to fill space quasiperiodically with rhombohedra, solids that are the three-dimensional equivalents of Penrose's diamond-shaped tiles. The two researchers ran computer simulations of such nonperiodic structures, showing the diffraction pattern that they would produce. The computer-generated diffraction patterns proved to be nearly identical to that of Shechtman's aluminum-manganese alloy.

That work provided important theoretical underpinning for Shechtman, who initially did not have an easy time convincing the scientific community that his discovery was the real thing. The quasiperiodic order in his diffraction pattern did not extend indefinitely. In fact, it extended for only a few hundred angstroms. One explanation immediately invoked by doubters was that the material was not a truly quasiperiodic crystal, but merely a set of crystallites arranged in an icosahedrally symmetric cluster, a phenomenon called twinning.

A growing crystal sometimes starts reproducing itself in one direction, producing a twin of itself. Shechtman's observations could potentially be explained away by assuming that twinning had created twins, which would deceptively appear to have the observed symmetry. It was not until 1984 that convincing arguments and further observations showed that the twinning model failed. Shechtman published his results, and shortly afterward, Steinhardt and Levine published their theory. Shechtman's experimental work and the Steinhardt-Levine theory supported each other well, however, and condensed matter physicists began to explore the realm of quasicrystals.

There followed a major burst of activity that led to the discovery and description of many more quasicrystalline materials. More than 100 quasicrystal alloys have now been identified. Each is made up of at least two different atomic species, with aluminum being one component in most cases. Most of these quasicrystals display icosahedral symmetry, but other symmetries have also been observed.

The first known quasicrystals were metastable. They were formed by rapid quenching of a liquid alloy, and if they were heated, their atoms were rearranged to form conventional metallic crystal structures. Later, alloys with stable quasicrystalline structures were discovered. The first example was an alloy of aluminum, lithium, and copper. These stable quasicrystals are valuable for research. Unlike the unstable materials, they can be solidified slowly from the liquid state, so that larger perfect quasicrystals can be formed, some of them centimeters in size. These quasicrystals make possible the measurement of physical properties such as thermal and electronic behavior.

Some doubts based on experimental evidence still remained, however.

Electron diffraction patterns showed the sharp Bragg peaks that indicate crystalline order. Diffraction patterns made with x rays or neutrons, however, had some fuzzy peaks, which could be interpreted as showing that quasicrystals did not fill space completely over very long ranges and that gaps would be found sooner or later if quasicrystals were made large enough. Two developments, one experimental and one theoretical, settled the issue in favor of quasicrystals.

The experimental evidence came from a Japanese research team that reported an alloy of aluminum, copper, and iron that had better icosahedral symmetry over longer distances than anything previously seen when examined by all methods of crystallography.

"When we looked at the new material, we found that we were getting Bragg peaks that were sharp and limited to the resolution of our high-resolution x-ray scattering apparatus, and these Bragg peaks were indexing exactly to an icosahedral reciprocal lattice," Bancel said. "This then was immediate evidence that we really had a quasiperiodic crystal and that these other models that had been brought forth could not be used to describe the aluminum-copper-iron phases," he explained.

Only a year earlier, there appeared to have been convincing arguments that it was impossible to grow such a perfect quasicrystal: to cover a surface with Penrose tiles, the tiles must be placed carefully in the precise order needed to ensure complete coverage; a slight error produces an incomplete configuration, with gaps in the coverage. Because of the long-range nature of the quasiperiodic order, critics expressed doubts that atoms could come together to grow perfect quasicrystals. An atom would have to sense the positions of other atoms at arbitrarily large distances to maintain the quasiperiodic order, and that seemed highly unphysical, the critics said. What in nature occurs that allows the three-dimensional analogs of Penrose tiles to fit together in a space-filling array?

A key turning point came in 1987, when it was realized that the argument based on Penrose tilings was fallacious, Steinhardt said. "Most importantly, it turned out that by making slightly longer-range rules than Penrose's matching rules you could avoid making mistakes altogether," he said. The new rules constrain the way in which two tiles are allowed to join not only along an edge but also around a vertex.

Penrose's rules required the tiles to fit together in specific ways based on matching faces, or sides, of the diamonds. Those rules, however, did not guarantee complete coverage of a surface. Someone who followed the rules could find a gap in the coverage after laying only a few tiles. In 1987, George Y. Onoda, a physicist at the

IBM Thomas J. Watson Research Center, working with Steinhardt and with David P. DiVincenzo of IBM and Joshua E.S. Socolar of Harvard University, developed a set of rules that led to a complete tiling scheme requiring only short-range interactions.

The original Penrose rules were based on marking edges of the two kinds of diamonds with different arrows pointing in different directions. A complete tiling under those rules required adjacent tiles to have matching arrow types and directions along shared edges. The new rules added two requirements, based on matching both edges and vertices of the Penrose tiles. The rules defined a "forced vertex" as one that was not completely surrounded by other tiles and had an edge that could be matched in only one way by another tile. One rule said that such new "forced tiles" should be added to a forced vertex until no forced vertices were left. The second rule said that in the absence of forced vertices, a fat diamond should be added to any corner of the array.

In computer simulations, the new rules produced infinitely large Penrose tilings, removing most of the theoretical objections. "The results provide new insights as to how materials with only short-range atomic interactions can grow large, nearly perfect quasicrystal grains," reported Onoda et al. (1988, p. 2653).

It is plausible that nature follows such rules. When a conventional crystal is grown from a seed, some sites of the crystal seed are "stickier" than others, encouraging atoms or molecules to attach there rather than at "nonsticky" sites. The same is true of quasicrystal seeds, Steinhardt said, with the sticky sites determined by a three-dimensional version of the Onoda et al. growth rules.

Assume a seed is immersed in a liquid containing atoms that will join the seed. In the case of a crystal, atoms join the seed at a regular rate, building layer after layer to form a large crystal over time. In the case of a perfect quasicrystal growing according to the rules of Onoda et al., the structure grows quickly at first, as atoms attach at the highly sticky forced sites. Occasionally, however, the cluster grows out to a surface that has no forced sites. Then, after some time, an atom eventually attaches at the less sticky corner sites. A quasicrystal thus displays "a herky-jerky growth process, every now and then waiting for the low-probability sticking sites to fill in when there are no high-sticking probability sites left," Steinhardt said. At the symposium, he showed a film of a computer simulation of a seed cluster of quasiperiodic tiles, representing atoms, in a liquid where it was bombarded constantly by other tiles. The film showed a stop-and-start growth process, with waits for a low-probability sticky site to be filled when all high-probability sticky sites were occupied. "This shows that if you allow me to choose the range of sticking probabilities, I can grow a quasicrystal to arbitrarily large finite size without making even one mistake," Steinhardt said. But the time needed to reach that size grows exponentially as the size increases, due to the less sticky sites, he added.

There is an important exception to that observation, however. One of the computer simulations run by Steinhardt started with a seed that had a specific defect, one that was discovered in the course of investigation. The defective seed looks like a ring of tiles with a hole in the center (Figure 10.2). The hole cannot be filled consistent with Penrose's rules, but the Penrose rules can be used to tile the outside of the ring and beyond. With this deliberately defective seed, growth never stops, because growth sites for the addition of atoms to the quasicrystal are always available. Steinhardt calls the defect that allows this quasicrystal growth a decapod because it has 10-fold symmetry.

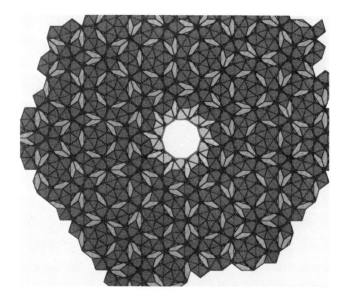

FIGURE 10.2 Beginning from the ring of tiles shown in the center, a Penrose tiling can be rapidly grown by attaching tiles only at highly sticky "force" sites. The ring of tiles that form the seed has a defect in the middle—the center cannot be tiled according to the proper Penrose matching rules. However, the outside can be tiled and grown without ever reaching a dead surface. (Courtesy of P. Steinhardt.)

Quasicrystals have provided a field day for mathematicians, who are familiar with quasiperiodic functions, which are the sum of periodic functions with periods of irrational ratio. A well-known example of a quasiperiodic function that relates to the new field of quasicrystals is the Fibonacci sequence, named for a 12th-century Italian mathematician. The Fibonacci sequence begins with 1, and each subsequent value is derived by taking the sum of the two preceding numbers; the first terms of the sequence are thus 1, 1, 2, 3, 5, 8, 13, 21, 34. . . . The ratio of two consecutive numbers approaches an irrational ratio called the golden ratio.

If the Penrose tiles are marked with specially chosen line segments, the segments join across tile edges to form five sets of infinite, parallel lines that are pentagonally oriented (Figure 10.3). Some of

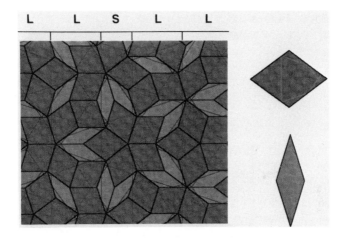

FIGURE 10.3 Penrose tiling showing the Ammann line decoration. The tiling consists of two types of tiles. Penrose discovered a set of matching rules that constrain the way two tiles can join together such that the only way to fill the plane consistent with those rules is by the Penrose tiling pattern above. Although there are many ways of choosing the tiles and matching rules, one way is to choose fat and skinny rhombus shapes and then to decorate each with the line segments shown on the right. (The lines are called Ammann lines because R. Ammann first suggested them.) The matching rule is that two tiles are allowed to join only if the segments join continuously across the interface. The segments join up to form a Fibonacci sequence of long and short intervals. (Courtesy of P. Steinhardt.)

the spaces between lines are short and some are long. The long-range spacing of short and long lines in each set of Ammann lines forms a Fibonacci sequence in which the ratio of long to short spacings approaches the golden ratio.

"It is not periodic, but it is ordered and predictable," Steinhardt said. "You can have both ordering and nonperiodicity."

Quasicrystals may have practical applications. There are indications that they may be highly resistant to deformation, which would make them valuable for use in heavy-duty bearings. The task of exploring their physical, as well as their mathematical, properties is just beginning.

"We're in the same boat as physicists were with crystals a hundred years ago," Steinhardt said: "We have the structure of quasicrystals. Now we must predict their electronic and physical properties. This is a mathematical challenge, because the mathematics for crystals doesn't work for quasicrystals."

BIBLIOGRAPHY

Onoda, George Y., Paul J. Steinhardt, David P. DiVincenzo, and Joshua E.S. Socolar. 1988. Growing perfect quasicrystals. Physical Review Letters 60:2653-2656.

RECOMMENDED READING

Asimov, Isaac. 1988. Understanding Physics. Dorset Press Reprint Series. Hippocrene Books, New York.

DiVincenzo, D., and P.J. Steinhardt (eds.). 1991. Quasicrystals: The State of the Art. World Scientific Publishing Company, Singapore.

Steinhardt, P.J. 1990. Quasicrystals: A New Form of Matter. Endeavour 14(3):112-116.

Steinhardt, P.J., and S. Ostlund (eds.). 1987. The Physics of Quasicrystals. World Scientific Publishing Company, Singapore.

Second Annual Frontiers of Science Symposium Program

Topic: ASTROPHYSICS

Margaret Geller and Larry Smarr, *Organizers*

SPEAKER: Tony Tyson, AT&T Bell Laboratories
TITLE: Imaging the Distant Universe: Galaxies and Dark
               Matter
DISCUSSANTS: S.G. Djorgovski, California Institute of Technology
               David Koo, University of California, Santa Cruz
               Edmund Bertschinger, Massachusetts Institute of
                  Technology
               Jill Bechtold, University of Arizona, Tucson

Topic: ATMOSPHERIC SCIENCE

Larry Smarr, *Organizer*

SPEAKER: Gregory McRae, Carnegie Mellon University
TITLE: Using Supercomputing and Visualization in Los
               Angeles Smog Simulation
DISCUSSANTS: Alan Lloyd, South Coast Air Quality Management
                  District, El Monte, California
               Arthur Winer, University of California, Los
                  Angeles

NOTE: Topics are listed in alphabetic order.

255

Topic: COMPUTATION

Larry Smarr, *Organizer*

SPEAKER: William Press, Harvard University
TITLE: Unexpected Scientific Computing
DISCUSSANTS: Jean Taylor, Rutgers University
Stephen Wolfram, Wolfram Research
Alan Huang, AT&T Bell Laboratories

Topic: DYNAMICAL SYSTEMS

William Thurston, *Organizer*

SPEAKER: John Hubbard, Cornell University
TITLE: Chaos, Determinism, and Information
DISCUSSANTS: Robert Devaney, Boston University
Steven Krantz, Washington University
Curt McMullen, University of California, Berkeley

Topic: GENE REGULATION

Eric Lander, *Organizer*

SPEAKER: Robert Tjian, University of California, Berkeley
TITLE: Gene Regulation in Animal Cells: Transcription
Factors and Mechanisms
DISCUSSANTS: Arnold Berk, University of California, Los Angeles
Douglas Hanahan, University of California, San
Francisco
Kevin Struhl, Harvard Medical School
Ruth Lehmann, Whitehead Institute for Biological
Research

Topic: GEOLOGY

Raymond Jeanloz and Susan Kieffer, *Organizers*

SPEAKER: David Stevenson, California Institute of
Technology
TITLE: How the Earth Works: Techniques for
Understanding the Dynamics and Structure of
Planets

DISCUSSANTS: Jeremy Bloxham, Harvard University
Michael Gurnis, University of Michigan
Russell Hemley, Carnegie Institution of
Washington
Marcia McNutt, Massachusetts Institute of
Technology

Topic: MAGNETIC RESONANCE IMAGING

Peter Dervan, *Organizer*

SPEAKER: William Bradley, Long Beach Memorial Medical
Center
TITLE: Clinical Magnetic Resonance Imaging and
Spectroscopy
DISCUSSANTS: David Stark, Massachusetts General Hospital
Robert Balaban, National Institutes of Health
Graeme Bydder, Royal Post-Graduate Medical
School, Hammersmith Hospital
John Crues III, Santa Barbara Cottage Hospital

Topic: NEURAL NETWORKS

Peter Dervan, *Organizer*

SPEAKER: Christof Koch, California Institute of Technology
TITLE: Visual Motion: From Computational Analysis
to Neural Networks and Perception
DISCUSSANTS: Paul Adams, Howard Hughes Medical Institute,
SUNY at Stony Brook
James Bower, California Institute of Technology
Terrence Sejnowski, Salk Institute
Shimon Ullman, Massachusetts Institute of
Technology

Topic: PHOTOSYNTHESIS

Peter Dervan, *Organizer*

SPEAKER: Mark Wrighton, Massachusetts Institute of
Technology
TITLE: Photosynthesis—Real and Artificial

DISCUSSANTS: Nathan Lewis, California Institute of Technology
Thomas Mallouk, University of Texas, Austin
George McLendon, University of Rochester
Douglas Rees, California Institute of Technology

Topic: PHYSICS

Orlando Alvarez, David Nelson, Raphael Kasper, *Organizers*

SPEAKER: Peter Bancel, IBM Corporation, Yorktown Heights
TITLE: Icosahedral Quasicrystals
SPEAKER: Alex Zettl, University of California, Berkeley
TITLE: High-Temperature Superconductivity
DISCUSSANTS: Daniel Fisher, Harvard University
Peter Gammel, AT&T Bell Laboratories
Duncan Haldane, University of California, La Jolla
Paul Steinhardt, University of Pennsylvania

Index